Math Mammoth
Add & Subtract 2-C
Elementary Math Workbook

By Maria Miller

ISBN 978-1-954379-44-2

2025 EDITION

Contents

Introduction

Math Mammoth Add and Subtract 2-C focuses on regrouping in addition (a.k.a. carrying) and in subtraction (a.k.a. borrowing), on a second grade level. The book assumes the student has already learned to add two-digit numbers with regrouping. In this book, students will learn 3-digit addition with regrouping, and both 2- and 3-digit subtraction with regrouping, with one exception: this book does not cover regrouping over zero tens in subtraction, as that topic is left for 3rd grade level in the Math Mammoth curriculum.

The free videos at **https://www.mathmammoth.com/videos/** (choose 2nd grade videos) can be used to help teach these topics.

In the first lesson, the student adds three-digit numbers with a regrouping in tens, but there is no regrouping in hundreds. Students already know how to regroup two-digit numbers, so this lesson simply extends that knowledge to numbers with three digits.

In the next lesson, students regroup 10 tens as a hundred (or carry to the hundreds). This is first illustrated with a visual model. You can adapt the exercises in the lesson to be done with manipulatives instead, if desired.

Then we study regrouping twice in addition: 10 ones form a new ten, and then 10 tens form a new hundred. Again, students first work with visual models, with the aim of helping them to understand the concept itself. Then, they learn the abstract process, adding the numbers in columns.

Next, we study regrouping in subtraction, starting with two-digit numbers. First, students are taught to break one ten into 10 ones. For example, 5 tens 4 ones is written as 4 tens 14 ones; one of the tens is "broken down" into 10 ones. This is the process of regrouping: one of the tens "changes groups" from being with the tens to being with the ones.

After mastering that process, it is time to use regrouping in subtraction problems and learn the traditional paper-and-pencil method of subtracting (where one number is written under the other).

Next, there is a lesson on word problems that include the thought of "more" or "fewer", and also several techniques or "tricks" for mental subtraction. The word problem sets in the book typically include both addition and subtraction problems. The reason for this is that students need to learn to recognize whether a problem requires addition or subtraction, and not blindly apply the operation mentioned in the title of the lesson.

After this, it is time to study regrouping in subtraction with three-digit numbers. There are four cases:

1. Regrouping one ten as 10 ones, such as is necessary in 546 − 229.
2. Regrouping one hundred as 10 tens, such as is necessary in 728 − 441.
3. Regrouping two times (one ten as 10 ones, and one hundred as 10 tens), such as in 725 − 448.
4. Regrouping with zero tens, such as is necessary in 405 − 278.

Since this book is geared towards second grade, we only study cases (1) and (2) from the list above. The other two cases are left for third grade; you can find lessons for those in our Blue Series book *Add & Subtract 3*).

Lastly, students work with bar graphs and play Euclid's game, which is meant as a fun, supplemental lesson.

I wish you success in teaching math!

Maria Miller, the author

Games and Activities

Missing Number Puzzles

Create puzzles for your student(s) by taking a simple addition or subtraction problem, and leaving out some of the digits. For example, the problem on the right is turned into a Missing Number Puzzle by leaving out three digits.

Original:

$$\begin{array}{r} 6\ 1\ 2 \\ -\ 1\ 7\ 1 \\ \hline 4\ 4\ 1 \end{array}$$

→

Puzzle form:

$$\begin{array}{r} \square\ 1\ \square \\ -\ 1\ \square\ 1 \\ \hline 4\ 4\ 1 \end{array}$$

For this chapter, use three-digit addition and both two and three-digit subtraction problems that involve regrouping.

You can also reverse the roles, and have your student make these types of puzzles for you to solve. Have your student check your work — and sometimes, make an intentional mistake for them to find!

This activity is from ***https://www.earlyfamilymath.org*** *and published here with permission.*

7-Card Draw to a Target

You need: Number cards from 0 through 9. (Standard playing cards work if you make, say, the queen to be zero. Or, play with numbers 1-9.) Paper and pencil for each player (for adding).

Game play: Choose a 3-digit target number, say 600. Each player takes seven cards from the deck, and uses those to form two 3-digit numbers to add (one card is left unused). Each player adds the two numbers they formed, using paper & pencil or mental math. The player closest to the target wins a point for that round.

The player with the highest number of points after, say, five rounds, wins.

This game is from ***https://www.earlyfamilymath.org*** *and published here with permission.*

Games and Activities at Math Mammoth Practice Zone

Color-Grid Game — Vertical Addition Practice
Solve 12 problems of adding three-digit numbers in columns.
https://www.mathmammoth.com/practice/vertical-addition#max=999&questions=4*3&addends=2&max-digits=3

Color-Grid Game — Vertical Subtraction Practice
Solve 12 problems of subtracting three-digit numbers in columns.
https://www.mathmammoth.com/practice/vertical-subtraction#max=999&zeros=0&questions=4*3

Hidden Picture Subtraction Game
Practice subtracting two-digit numbers with mental math, and reveal a hidden picture!
https://www.mathmammoth.com/practice/mystery-picture-subtraction#min=11&max=99

Two-Digit Subtraction with Mental Math
Simple online practice of subtracting two-digit numbers using mental math.

- Subtract a single-digit number from a two-digit number:
  https://www.mathmammoth.com/practice/addition-subtraction-two-digit#opts=2m1dwr&questions=10
- Subtract 2-digit numbers:
  https://www.mathmammoth.com/practice/addition-subtraction-two-digit#opts=2m2dwr&questions=10

Mathy's Berry Picking Adventure
Practice subtracting two-digit numbers (e.g. 52 − 17).
https://www.mathmammoth.com/practice/mathy-berries#mode=sub-100&duration=2m

Bingo
Choose Subtraction (Two-Digit) to practice mental subtraction of two-digit numbers.
https://www.mathmammoth.com/practice/bingo

Fruity Math — two-digit subtraction
Click the fruit with the correct answer and try to get as many points as you can within two minutes.
https://www.mathmammoth.com/practice/fruity-math#op=subtraction&duration=60&mode=manual&config=21,99x1__6,80x1&allow-neg=0

Make number sentences
Drag two flowers to the empty slots to make the given difference, practicing two-digit mental subtraction.
https://www.mathmammoth.com/practice/number-sentences#questions=5&types=sub-6-100

Further Resources on the Internet

We have compiled a list of external Internet resources that match the topics in this book. This list of links includes web pages that offer:

- **online practice** for concepts;
- online **games**, or occasionally, printable games;
- **animations** and interactive **illustrations** of math concepts;
- **articles** that teach a math concept.

We heartily recommend you take a look at the list. Many of our customers love using these resources to supplement the bookwork. You can use the resources as you see fit for extra practice, to illustrate a concept better, and even just for some fun. Enjoy!

https://l.mathmammoth.com/blue/addsubtract2C

Scan me

Adding 3-Digit Numbers in Columns

When the numbers have three digits, adding them in columns (where you write the numbers under each other) is still easy. We just add hundreds in their own column.

	hundreds	tens	ones
	2	4	1
+	3	3	5
	5	7	6

Regrouping in the tens happens the same way as what you have learned before.

Ten ones form a new ten. This new ten is regrouped with the other tens. We write a little "1" above the other tens in the tens' column.

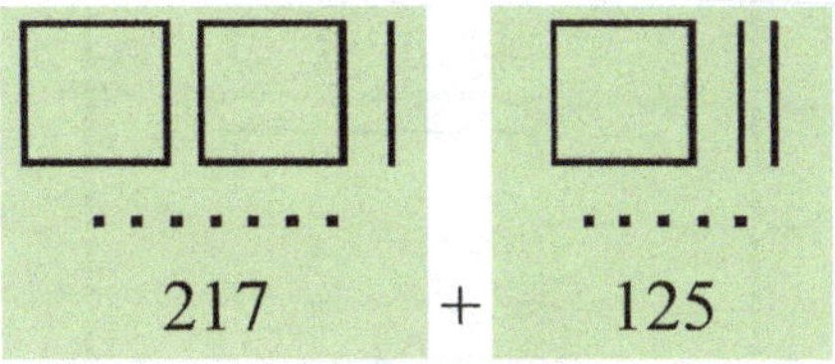

Circle ten ones to make a new ten!

	hundreds	tens	ones
		1	
	2	1	7
+	1	2	5
	3	4	2

1. Write the numbers to be added in the grid, and add. Regroup.

a.

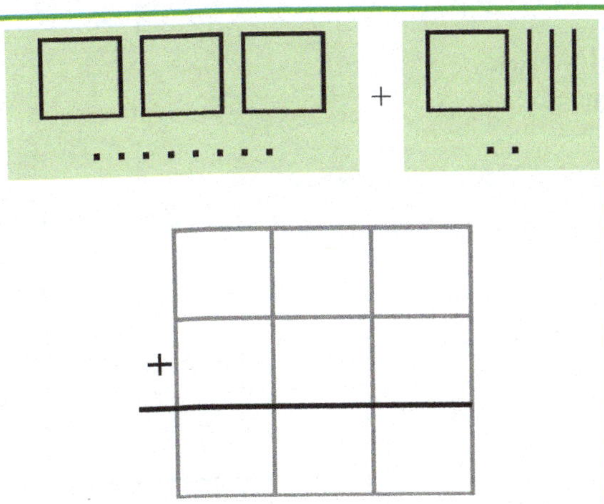

b.

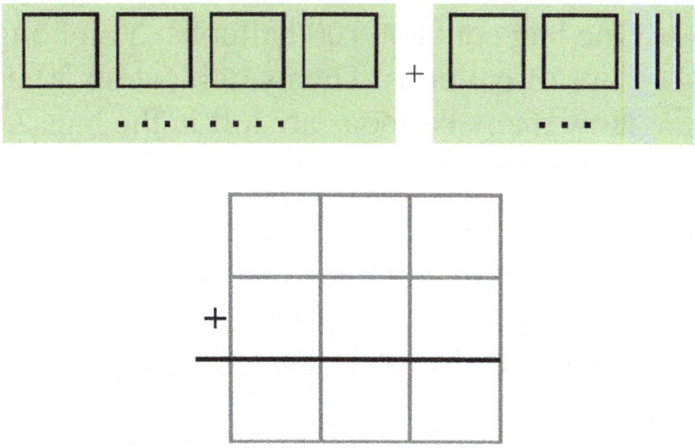

c.

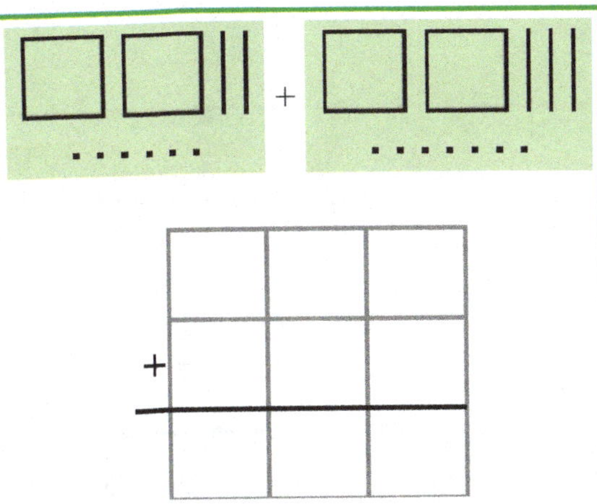

d.

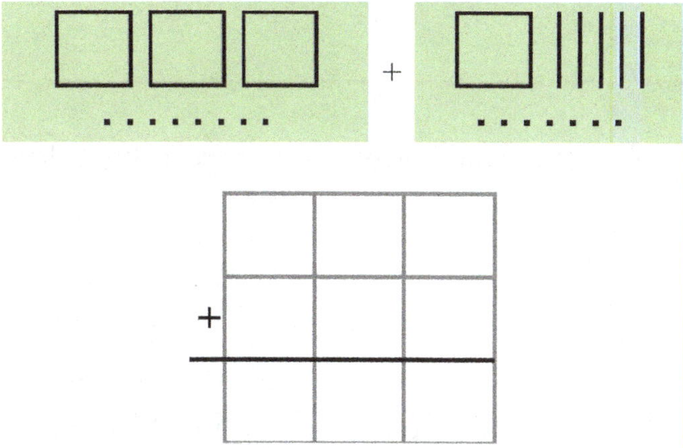

2. Add. Regroup 10 ones as a ten if necessary. Add hundreds in their own column.

a. 3 0 1 + 5 1 0	**b.** 2 2 7 + 5 3 6	**c.** 4 0 2 + 3 7 0	**d.** 1 2 9 + 4 5 4
e. 2 0 7 2 2 3 + 4 3 6	**f.** 4 3 7 1 1 5 + 4 3 7	**g.** 2 2 4 3 0 1 + 4 0 8	**h.** 2 3 2 3 0 8 + 4 2 5

3. **a.** A factory had 547 workers. Then the factory hired 128 more workers. How many workers are there now?

+		

b. One bag contains 100 balloons. You bought six bags of balloons. Then you blew up 20 balloons. How many balloons are left in the bags?

c. Nancy bought three bookcases. Each bookcase cost $116. What was the total cost?

+		

Puzzle Corner

What numbers are missing from the additions?

□ 3 □ + 1 □ 5 7 7 6	2 □ □ + □ 5 5 5 8 0	1 1 6 + □ 5 □ 5 □ 3	□ 4 7 + 7 □ □ 9 9 2

Regrouping 10 Tens as a Hundred

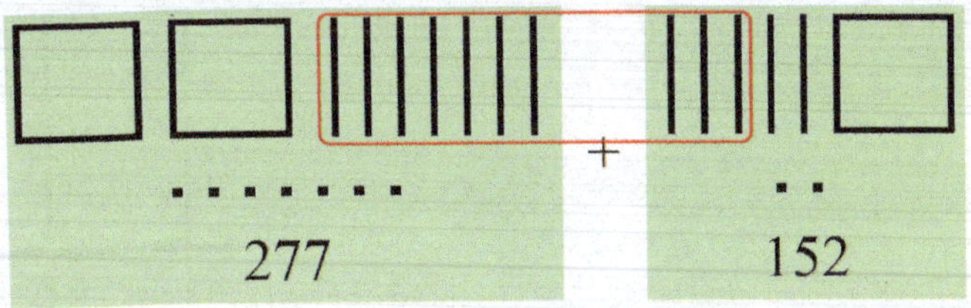

hundreds	tens	ones
1		
2	7	7
+ 1	5	2
4	2	9

Ten TENS are grouped to make a new hundred!

In the TENS, there are 7 tens and 5 tens to add (277 has 7 tens, and 152 has 5 tens).

That is 12 tens. And, **10 tens makes a hundred!** So, we make a new hundred, and regroup that with the other hundreds, writing the new hundred with a little "1" in the hundreds column.

(We have 2 tens left over from that, and they stay in the tens column.)

1. Circle ten 10-sticks to make a new hundred. Write the addition.
   Alternatively, you can do these exercises using base-ten blocks or similar manipulatives.

a. 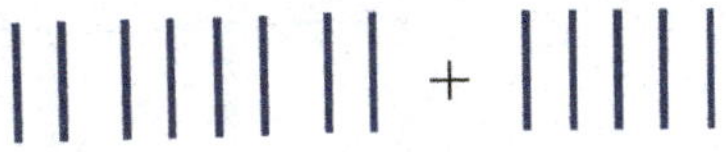

_______ + _______ = _______

b. 

_______ + _______ = _______

c. 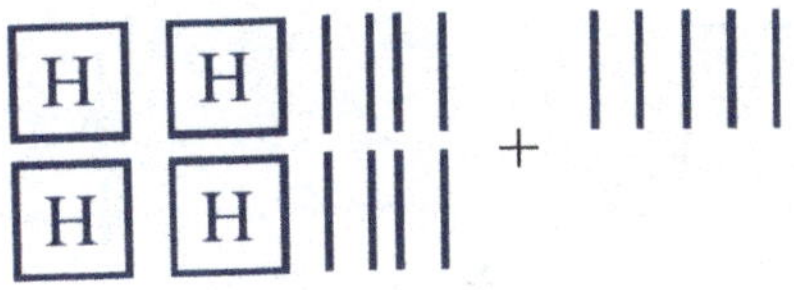

_______ + _______ = _______

d. 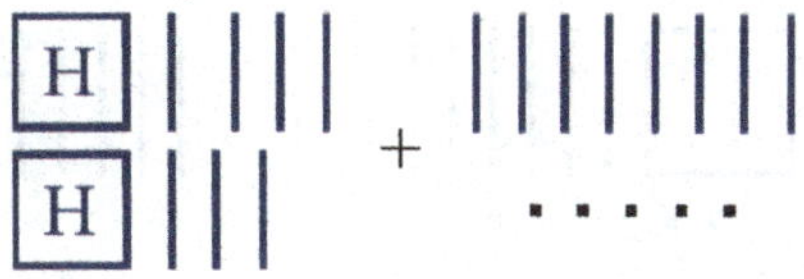

_______ + _______ = _______

e. 

_______ + _______ = _______

f. 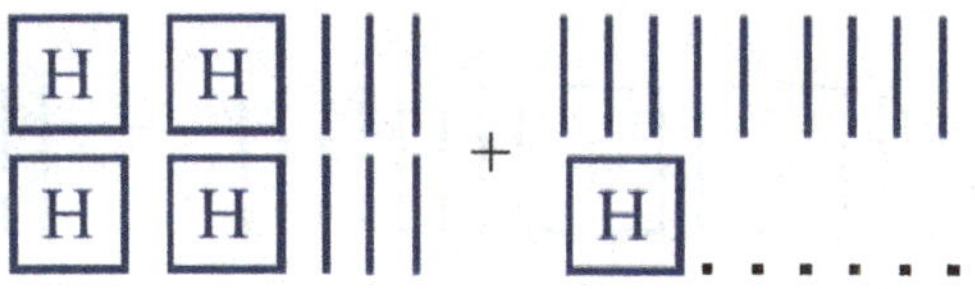

_______ + _______ = _______

2. Write the numbers in the grid, and add. Regroup. You can circle 10 ten-sticks in the picture to help you. Alternatively, you can do these exercises using base-ten blocks or similar manipulatives.

a. 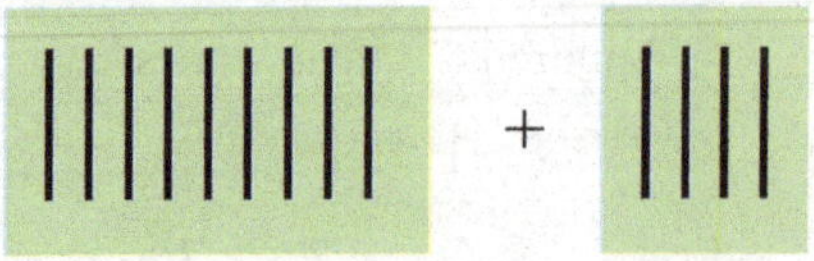

90 + 40

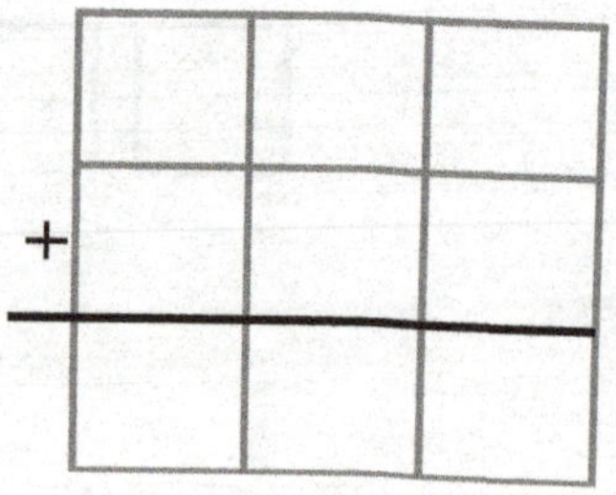

b. 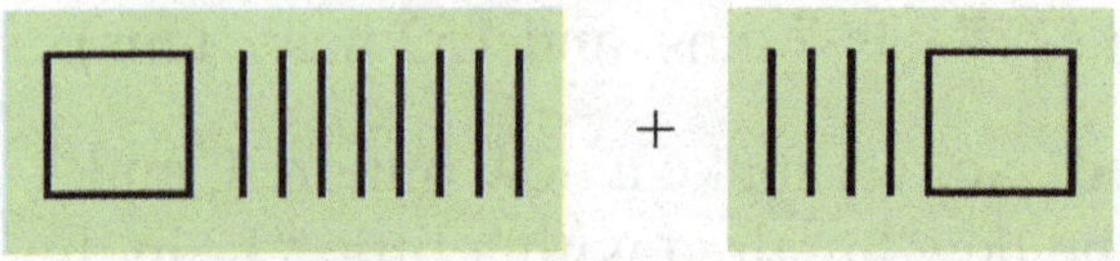

180 + 140

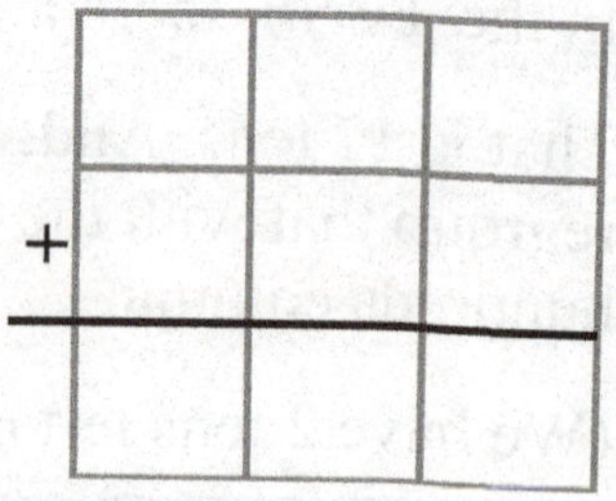

c. 

350 + 63

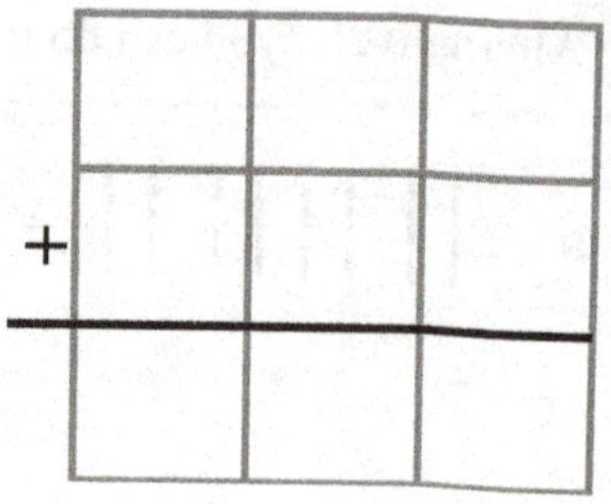

d. 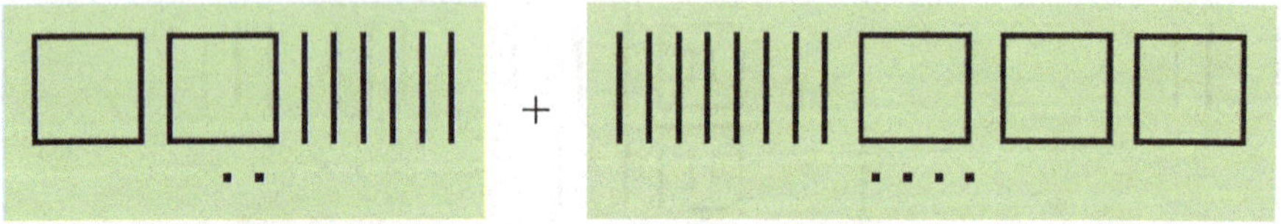

262 + 384

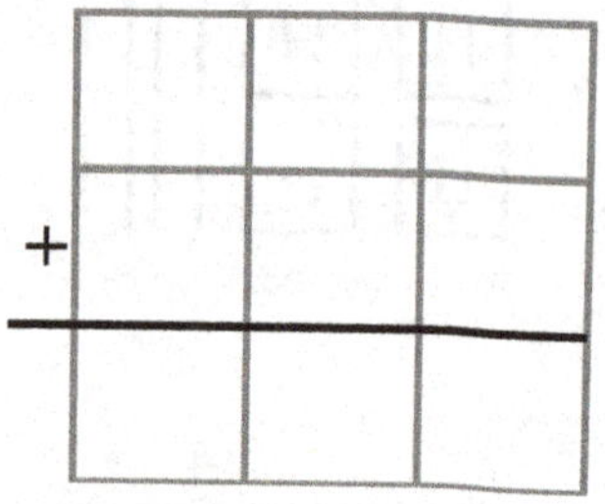

e. 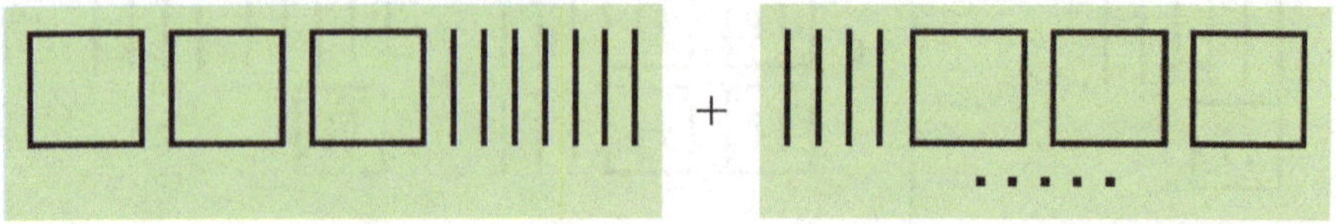

370 + 345

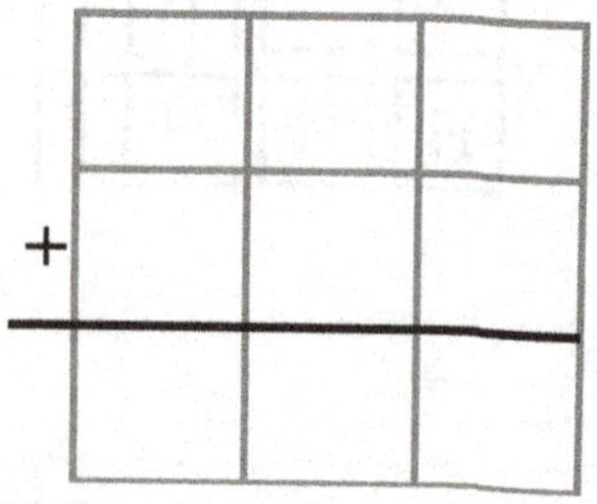

3. Add mentally. Compare the problems. Notice you get a NEW hundred from the tens!

a.	b.	c.
70 + 40 = ________	50 + 60 = ________	90 + 50 = ________
170 + 40 = ________	150 + 60 = ________	290 + 50 = ________
270 + 40 = ________	250 + 60 = ________	490 + 50 = ________

4. Add. You need to regroup 10 tens as a new hundred.

a. 80 + 30 = ____

b. 220 + 90 = ____

c. 64 + 53 = ____

d. 370 + 74 = ____

e. 533 + 282 = ____

f. 67 + 72 = ____

g. 224 + 193 = ____

h. 464 + 392 = ____

i. 355 + 374 = ____

j. 787 + 82 = ____

5. Add mentally. THINK of the new hundred you might get from adding the tens.

a.	b.	c.
70 + 40 = ________	80 + 60 = ________	290 + 50 = ________
130 + 40 = ________	270 + 60 = ________	220 + 50 = ________
160 + 50 = ________	130 + 50 = ________	190 + 20 = ________

6. What number was added? Think of regrouping!

a. 167 + 1__2 = 359

b. 240 + 1__2 = 422

c. 391 + 4__2 = 813

d. 653 + 1__3 = 846

e. 375 + 1__4 = 559

7. Add and match the letters with the answers in the key. The key then solves the riddle.

W	L	P	T	S
233 + 758	553 + 346	597 + 330	191 + 751	282 + 647
E	**O**	**A**	**E**	**I**
111 + 729	772 + 132	474 + 343	217 + 639	470 + 399
G	**N**	**R**	**F**	**H**
216 116 + 529	231 240 + 432	85 205 + 643	136 134 + 589	105 301 + 459

Key:

817	840	856	859	861	865	869	899	903	904	927	933	929	942	991

1. How do you put an elephant in the fridge?
   - You open the door, put the elephant in, and close the door. ☺

2. How do you put a giraffe in the fridge?
   - You open the door, take the elephant out, put the giraffe in, and close the door. ☺

3. When the elephant and the giraffe ran a race, who won?

942	865	840

856	899	856	927	865	817	903	942

, because

942	865	856

861	869	933	817	859	859	840

991	817	929

869	903

942	865	856

933	840	859	933	869	861	856	933	817	942	904	933

Add in Columns: Regrouping Twice

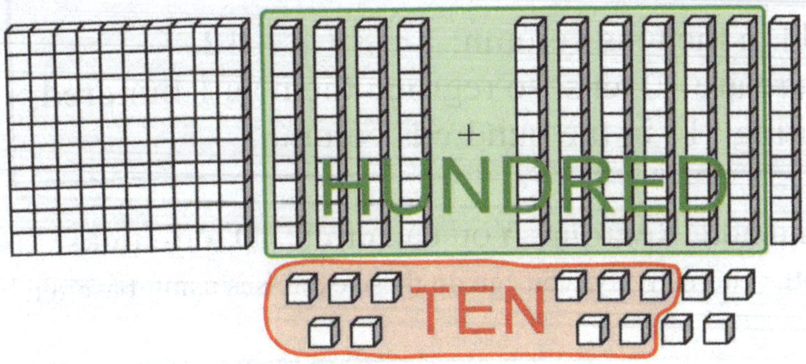

145 + 79

10 ones form a new ten. 10 tens form a new hundred.

The total is 224. Can you see that in the picture?

(You can also use manipulatives to do this problem.)

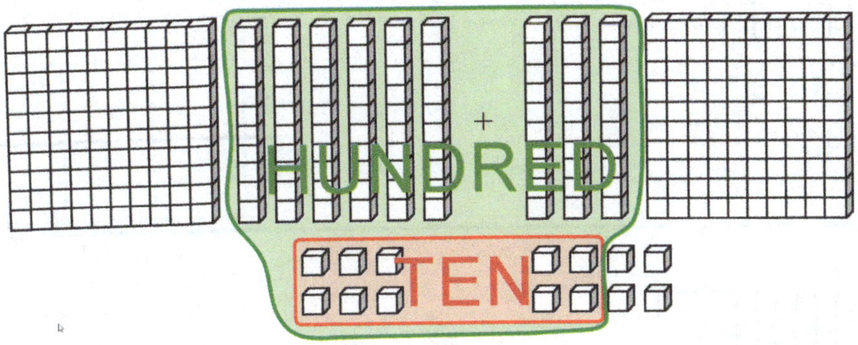

166 + 138

10 ones form a new ten. 10 tens form a new hundred.

The total is 304. Can you see that in the picture?

(You can also use manipulatives to do this problem.)

You have to regroup the ones and the tens. You have to regroup two times.

1. Circle ten 1-dots to make a new ten, AND circle ten 10-sticks to make a new hundred. Write the addition. Alternatively, you can do these exercises using base-ten blocks or similar manipulatives.

a. 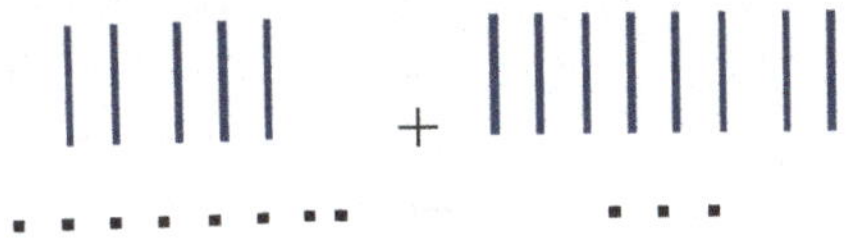

________ + ________ = ________

b. 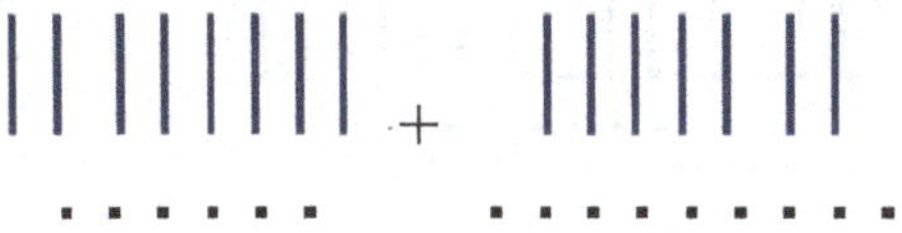

________ + ________ = ________

c. 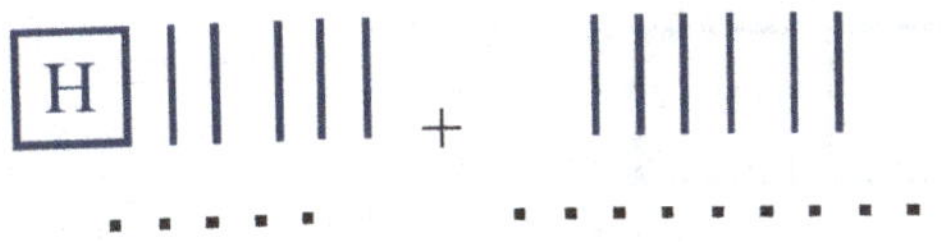

________ + ________ = ________

d. 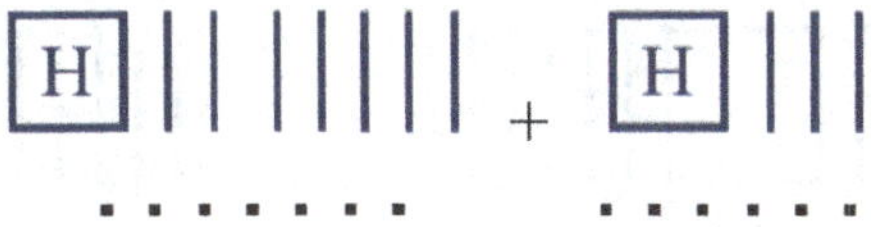

________ + ________ = ________

hundreds	tens	ones
1	1	
1	8	7
+ 1	3	8
3	2	5

Add in the ones column: 7 + 8 = 15.
There are more than 10 ones, so regroup them as 1 ten 5 ones, writing "1" in the tens column.

Add in the tens column: 1 + 8 + 3 = 12.
There are 10 tens, so regroup them as 1 hundred, writing "1" in the hundreds column.

2. Write the numbers in the grid, and add. Regroup. You can circle 10 ten-sticks AND 10 ones in the picture to help you. Alternatively, you can do these exercises using base-ten blocks or similar manipulatives.

a.  + 

96 + 145

b.  + 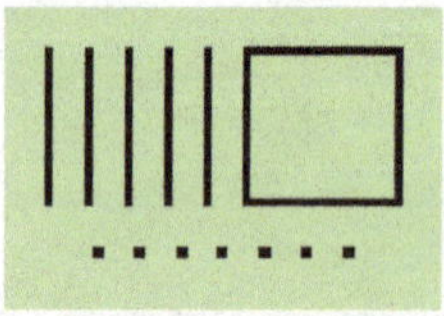

185 + 157

c. 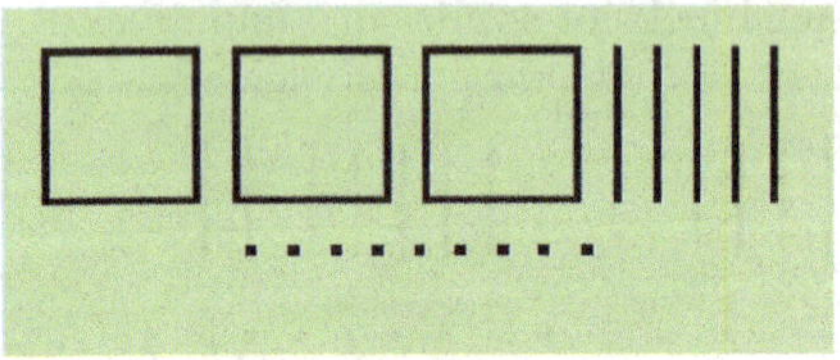 + 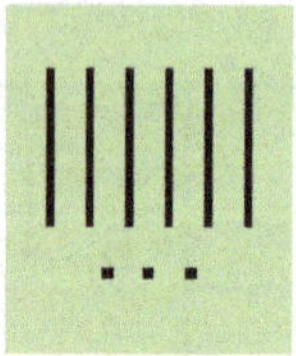

359 + 63

d. 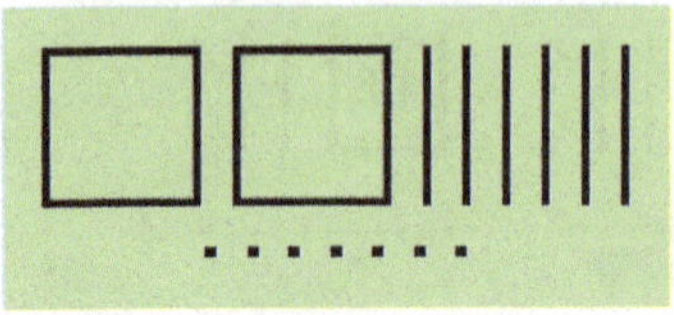 + 

267 + 385

3. Mary added 256 and 384 using the picture. Explain how she did it.

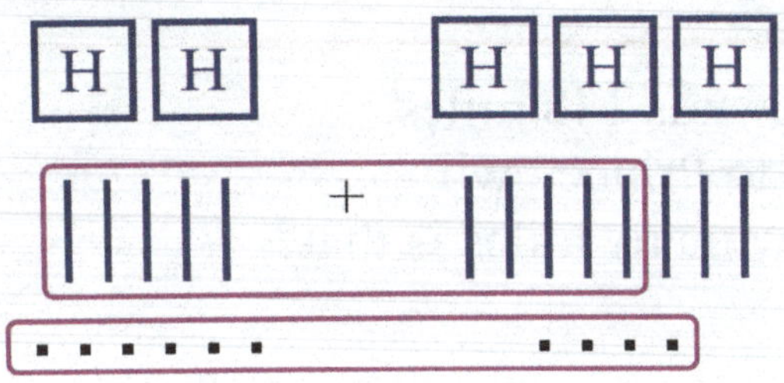

256 + 384 = 6 hundreds and 4 tens = 640

4. Add. Regroup two times if necessary.

a. 306 + 461	b. 299 + 225	c. 488 + 322	d. 115 + 536
e. 704 + 156	f. 260 + 341	g. 248 + 376	h. 173 + 646
i. 404 + 199 + 156	j. 701 + 129 + 101	k. 335 + 219 + 278	l. 103 + 280 + 547

5. Matt solved 650 + 331 in an interesting way. Can you follow his thinking? Fill in.

First I check the hundreds: 600 + 300 makes __________.

Then I add the ______________, and I get 50 + _______ = _______.

Lastly in the ones, there is just 0 and 1, which is 1.

Okay, so I have these parts: 900, 80, and _____, so that makes _________________.

6. Solve the word problems.

a. From Flowertown to Princetown is 148 miles. You travel from Flowertown to Princetown and back to Flowertown. How many miles is that?

b. The school bought pencils for $128, pens for $219, and notebooks for $549. Find the total cost of the items.

+

c. Find how many feet it is if you walk all of the way around this triangle.

+

Puzzle Corner What numbers are missing from the addition problems?

□ 3 □	2 □ □	1 6 9	□ 8 8
+ 1 □ 9	+ □ 3 6	+ □ 5 □	+ 7 □ □
3 9 1	5 1 7	7 □ 4	9 0 0

Regrouping in Subtraction, Part 1

We will now study regrouping ("borrowing") in subtraction.

As a first step, we study breaking a ten-pillar into ten little cubes. This is called ***regrouping***, because one ten "changes groups" from the tens group into the ones.

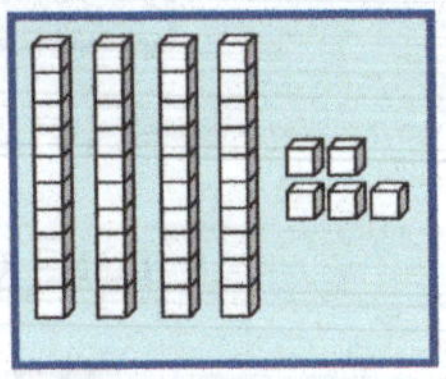

4 tens 5 ones

First we have 45. We "break" one ten-pillar into little cubes.

Break a ten. ⇨

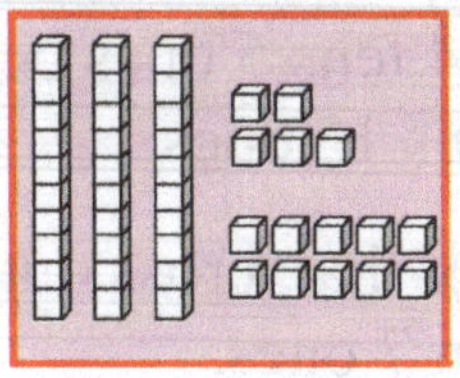

3 tens 15 ones

Now we have 3 tens and 15 ones. It is still 45, but written in a different way.

Here is another example. First we have 5 tens 3 ones. We "break" one ten-pillar into 10 little cubes. We end up with 4 tens 13 ones.

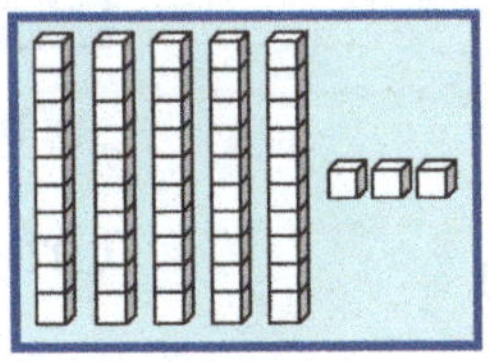

5 tens 3 ones

Break a ten. ⇨

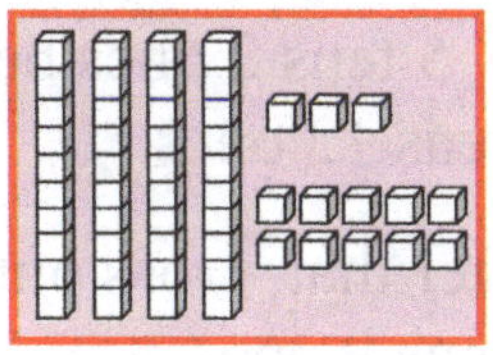

4 tens 13 ones

1. Break a ten into 10 ones. What do you get? Draw or use manipulatives to help.

a. 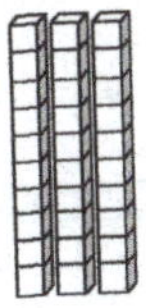 ⇨

3 tens 0 ones ⇨ ___tens ____ones

b. 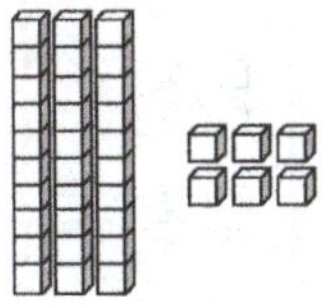 ⇨

___tens ___ones ⇨ ___tens ____ones

c. 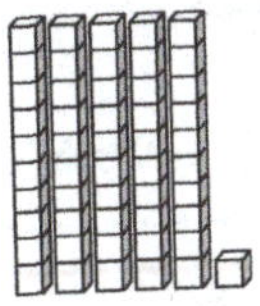 ⇨

___tens ___ones ⇨ ___tens ____ones

d. 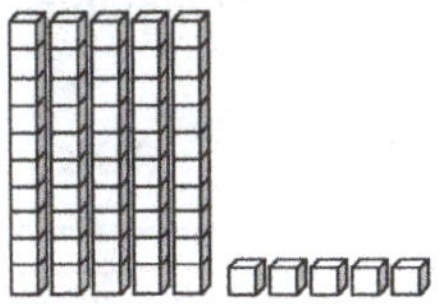 ⇨

___tens ___ones ⇨ ___tens ____ones

e. 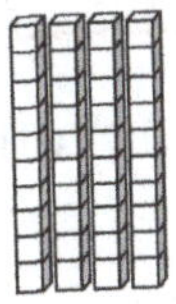 ⇨

___tens ___ones ⇨ ___tens ____ones

f. 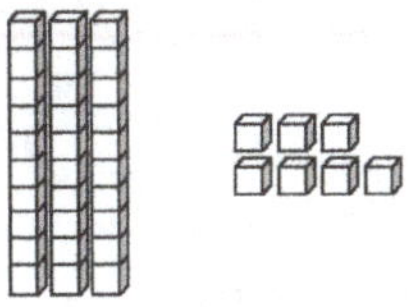 ⇨

___tens ___ones ⇨ ___tens ____ones

Let's study subtraction. The pictures on the right illustrate 45 – 17.

First, a ten is broken into 10 ones. So, 4 tens 5 ones becomes 3 tens 15 ones.

After that, cross out (subtract) 1 ten 7 ones.

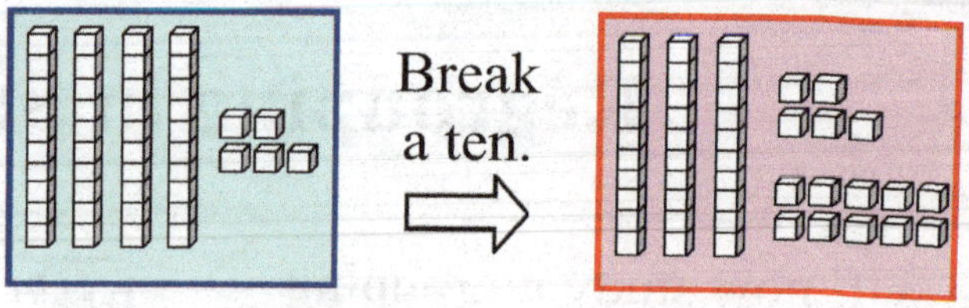

4 tens 5 ones   3 tens 15 ones

Cross out 1 ten 7 ones (from the *second* picture).

What is left? ____ tens ____ ones

The pictures on the right illustrate 52 – 39.

First, a ten is broken into 10 ones. So, 5 tens 2 ones becomes 4 tens 12 ones.

After that, cross out (subtract) 3 tens 9 ones.

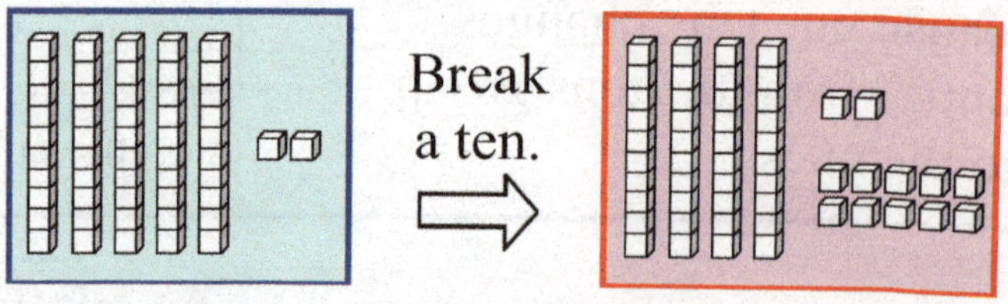

5 tens 2 ones   4 tens 12 ones

Cross out 3 tens 9 ones (from the *second* picture).

What is left? ____ tens ____ ones

2. Fill in. Always subtract (cross out some) from the *second* picture.

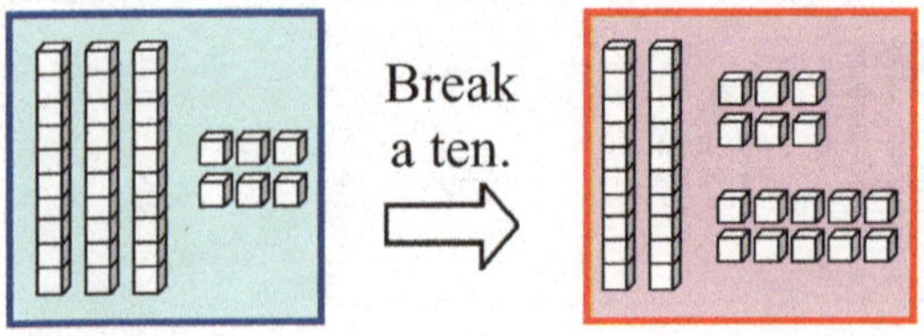

3 tens 6 ones   2 tens 16 ones

a. Subtract 8 ones (from the *second* picture).

What is left? ____ tens ____ ones

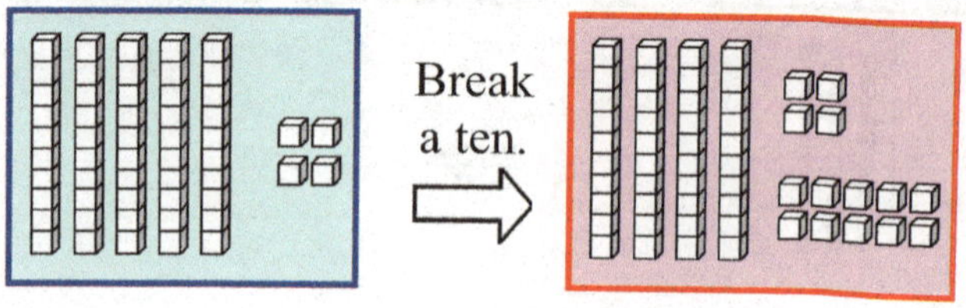

___ tens ___ ones   ___ tens ___ ones

b. Subtract 2 tens 7 ones.

What is left? ____ tens ____ ones

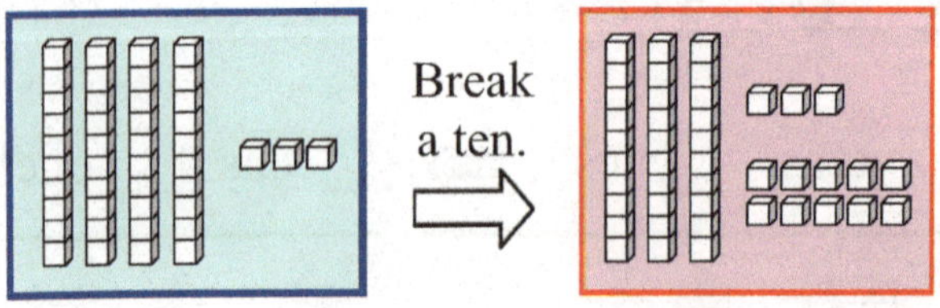

___ tens ___ ones   ___ tens ___ ones

c. Cross out 2 tens 5 ones.

What is left? ____ tens ____ ones

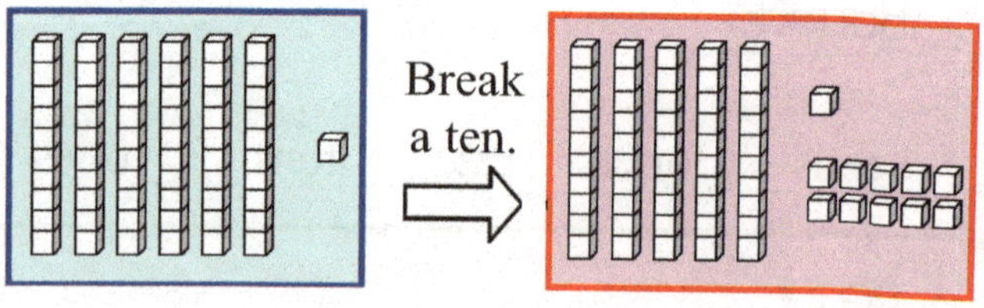

___ tens ___ ones   ___ tens ___ ones

d. Cross out 4 tens 4 ones.

What is left? ____ tens ____ ones

3. First, break a ten. Then subtract ones and tens separately. Look at the example.

a. 5 tens 5 ones ⇨ 4 tens 15 ones − 3 tens 7 ones 1 ten 8 ones	**b.** 7 tens 2 ones ⇨ ___ tens ___ ones − 3 tens 5 ones ___ tens ___ ones
c. 6 tens 0 ones ⇨ ___ tens ___ ones − 2 tens 7 ones ___ tens ___ ones	**d.** 6 tens 4 ones ⇨ ___ tens ___ ones − 3 tens 8 ones ___ tens ___ ones
e. 7 tens 6 ones ⇨ ___ tens ___ ones − 4 tens 7 ones ___ tens ___ ones	**f.** 5 tens 0 ones ⇨ ___ tens ___ ones − 2 tens 2 ones ___ tens ___ ones
g. 8 tens 1 one ⇨ ___ tens ___ ones − 6 tens 5 ones ___ tens ___ ones	**h.** 6 tens 3 ones ⇨ ___ tens ___ ones − 2 tens 8 ones ___ tens ___ ones

4. Jessica had 27 colored pencils and her brother and sister had none. Then Jessica gave 10 of them to her brother, and four to her sister.

a. How many pencils does Jessica have now?

b. How many more pencils does Jessica have than her brother?

c. How many more pencils does Jessica have than her sister?

Regrouping in Subtraction, Part 2

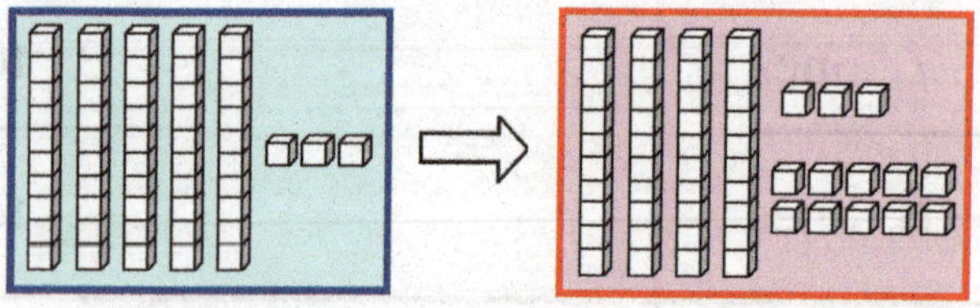

5 tens, 3 ones → 4 tens, 13 ones

In columns:

4	13
~~5~~	~~3~~
− 1	6
3	7

Cross out 1 ten 6 ones. What is left? ____ tens ____ ones

When the subtraction is done *in columns*:

- We take (borrow) one ten from the 5 tens. There will be now only 4 tens in the tens column, so to show this, we **cross the "5"** in the tens column and **write 4 above it**.
- This new ten is now grouped with the ones. There were 3 ones, but with the 10 new ones there will be 13. To show this, we also **cross the "3"** in the ones column and **write 13 above it**.
- Then we subtract the tens and ones separately.

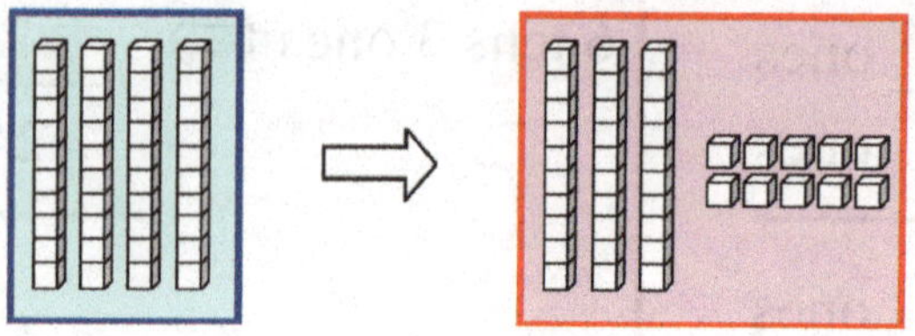

4 tens → 3 tens, 10 ones

3	10
4	~~0~~
− 2	8
1	2

Cross out 2 tens 8 ones. What is left? ___ tens ___ ones

Here is another example: 40 – 28.

- We take (borrow) one ten from the 4 tens. There will be now only 3 tens in the tens column, so to show this, we **cross the "4"** in the tens column and **write 3 above it**.
- This new ten is now grouped with the ones. There were 0 ones, but with the 10 new ones there will be 10. To show this, we also **cross the "0"** in the ones column and **write 10 above it**.
- Then we subtract the tens and ones separately in columns.

1. Regroup first. Then subtract.

a. 6 tens 0 ones → ____ tens ____ ones

Take away 3 tens, 9 ones.

	6	0
−	3	9

b. 7 tens 1 one → ____ tens ____ ones

Take away 1 ten, 6 ones.

	7	1
−	1	6

c. 3 tens, 5 ones → ____ tens ____ ones

Take away 1 ten, 7 ones.

	3	5
−	1	7

d. 8 tens → ____ tens ____ ones

Take away 3 tens, 4 ones.

	8	0
−	3	4

e. 7 tens, 6 ones → ____ tens ____ ones

Take away 4 tens, 8 ones.

	7	6
−	4	8

f. 9 tens → ____ tens ____ ones

Take away 5 tens, 1 one.

	9	0
−	5	1

g. 5 tens, 4 ones → ____ tens ____ ones

Take away 2 tens, 5 ones.

−		

h. 8 tens → ____ tens ____ ones

Take away 4 tens, 7 ones.

−		

i. 7 tens, 4 ones → ____ tens ____ ones

Take away 3 tens, 8 ones.

−		

j. 4 tens 7 ones → ____ tens ____ ones

Take away 2 tens, 9 ones.

−		

2. Subtract. Check by adding the result and what was subtracted.

a. $\begin{array}{r} \overset{4}{\cancel{5}}\,\overset{16}{\cancel{6}} \\ -\ 2\ 7 \\ \hline 2\ 9 \end{array}$ Check: $\begin{array}{r} \overset{1}{2}\ 9 \\ +\ 2\ 7 \\ \hline 5\ 6 \end{array}$	**b.** $\begin{array}{r} 9\ 0 \\ -\ 2\ 8 \\ \hline \end{array}$ Check: $\begin{array}{r} \\ +\ 2\ 8 \\ \hline \end{array}$	**c.** $\begin{array}{r} 4\ 2 \\ -\ 1\ 5 \\ \hline \end{array}$ Check: $\begin{array}{r} \\ +\ 1\ 5 \\ \hline \end{array}$
d. $\begin{array}{r} 9\ 0 \\ -\ 3\ 5 \\ \hline \end{array}$	**e.** $\begin{array}{r} 8\ 2 \\ -\ 2\ 5 \\ \hline \end{array}$	**f.** $\begin{array}{r} 6\ 5 \\ -\ 3\ 9 \\ \hline \end{array}$
g. $\begin{array}{r} 5\ 2 \\ -\ 1\ 4 \\ \hline \end{array}$	**h.** $\begin{array}{r} 6\ 5 \\ -\ 2\ 6 \\ \hline \end{array}$	**i.** $\begin{array}{r} 7\ 0 \\ -\ 4\ 8 \\ \hline \end{array}$
j. $\begin{array}{r} 5\ 5 \\ -\ 1\ 7 \\ \hline \end{array}$	**k.** $\begin{array}{r} 3\ 1 \\ -\ 1\ 8 \\ \hline \end{array}$	**l.** $\begin{array}{r} 6\ 6 \\ -\ 2\ 8 \\ \hline \end{array}$

Puzzle Corner

Find the missing numbers in these subtractions.
You might need to regroup.

$\begin{array}{r} \square\ 3 \\ -\ 1\ \square \\ \hline 7\ 5 \end{array}$ $\quad$ $\begin{array}{r} 8\ \square \\ -\ \square\ 7 \\ \hline 1\ 6 \end{array}$ $\quad$ $\begin{array}{r} \square\ 0 \\ -\ 3\ \square \\ \hline 4\ 2 \end{array}$ $\quad$ $\begin{array}{r} \square\ \square \\ -\ 1\ 4 \\ \hline 6\ 8 \end{array}$ $\quad$ $\begin{array}{r} 6\ 2 \\ -\ \square\ \square \\ \hline 5\ 3 \end{array}$

Regrouping in Subtraction, Part 3

Sometimes we need to regroup in subtraction, and sometimes not. Check carefully in the ones column. Are there enough ones to do the subtraction, or not? If not, you need to regroup.	Do you need to regroup? YES / NO $\begin{array}{r} 61 \\ -\ 26 \\ \hline \end{array}$	Do you need to regroup? YES / NO $\begin{array}{r} 74 \\ -\ 23 \\ \hline \end{array}$

1. Look at the ones' digits. Do you need to regroup (borrow a ten in the ones' column)?

a. Do you need to regroup? YES / NO $\begin{array}{r} 54 \\ -\ 32 \\ \hline \end{array}$	**b.** Do you need to regroup? YES / NO $\begin{array}{r} 50 \\ -\ 25 \\ \hline \end{array}$	**c.** Do you need to regroup? YES / NO $\begin{array}{r} 82 \\ -\ 56 \\ \hline \end{array}$

2. Subtract. Regroup if necessary. Find the answers in the line of numbers below.

a. Do you need to regroup? YES / NO $\begin{array}{r} 60 \\ -\ 16 \\ \hline \end{array}$	**b.** Do you need to regroup? YES / NO $\begin{array}{r} 57 \\ -\ 32 \\ \hline \end{array}$	**c.** Do you need to regroup? YES / NO $\begin{array}{r} 43 \\ -\ 17 \\ \hline \end{array}$

d. $\begin{array}{r} 80 \\ -\ 28 \\ \hline \end{array}$	**e.** $\begin{array}{r} 97 \\ -\ 25 \\ \hline \end{array}$	**f.** $\begin{array}{r} 81 \\ -\ 57 \\ \hline \end{array}$	**g.** $\begin{array}{r} 37 \\ -\ 27 \\ \hline \end{array}$	**h.** $\begin{array}{r} 60 \\ -\ 41 \\ \hline \end{array}$

44 26 19 25 24 72 10 52 25

3. Subtract mentally.

a. 64 – 20 = ______ 98 – 50 = ______	**b.** 77 – 71 = ______ 45 – 40 = ______	**c.** 98 – 6 = ______ 50 – 46 = ______

The number line arrows illustrate the subtraction 23 – 7. The first, red, arrow goes from 0 to 23. The second, blue, arrow goes 7 steps backwards from 23 and ends at 16.

4. Write the subtractions illustrated by the arrows on the number line.

a.
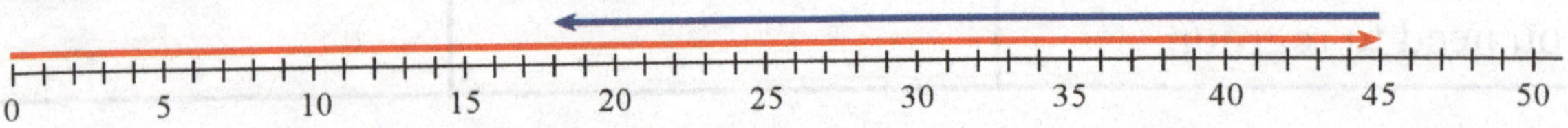

b.
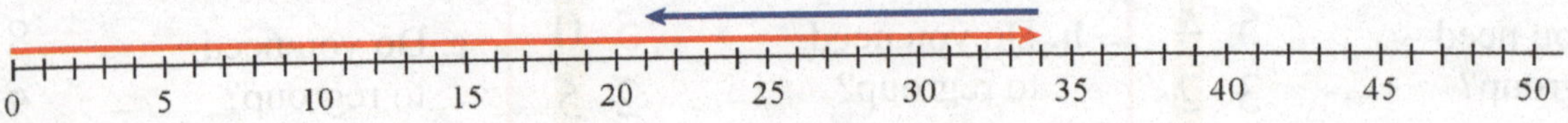

c.
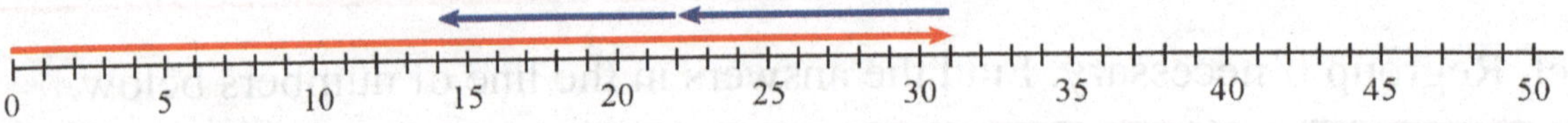

5. Draw arrows to illustrate these subtractions on the number line.

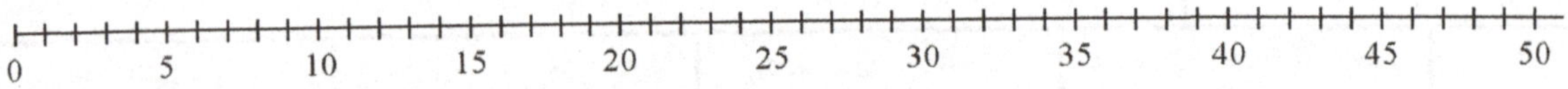

a. 22 – 9 = _____

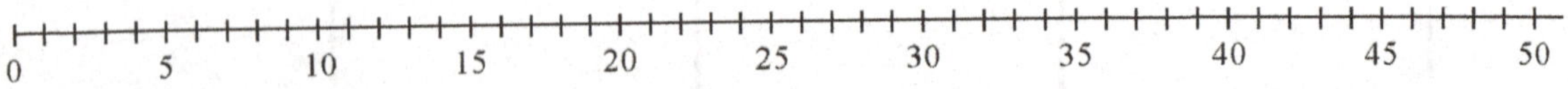

b. 36 – 12 = _____

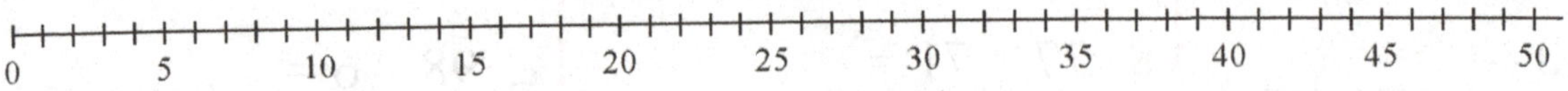

c. 44 – 17 = _____

Remember, subtraction and addition are connected. For example, 9 – 4 = 5 and 5 + 4 = 9. You can use this connection, and <u>check</u> each subtraction by adding. Add the answer you got *and* the number you subtracted. You should get the number you subtracted from. For example, to check if 68 – 45 is really 23, add 23 + 45 and check if you get 68.	4 14 ~~5~~ ~~4~~ – 3 7 ——— 1 7 1 1 7 + 3 7 ——— 5 4

6. Subtract. Regroup if necessary. Check each subtraction by *adding your answer and the number you subtracted.*

a. 8 14 ~~9~~ 4 – 3 5 ——— 5 9 5 9 + 3 5 ———	**b.** 8 2 – 2 5 ——— + 2 5 ———
c. 6 1 – 4 9 ——— + 4 9 ———	**d.** 9 9 – 5 7 ——— + 5 7 ———
e. 6 0 – 2 3 ——— + 2 3 ———	**f.** 6 6 – 4 8 ——— + 4 8 ———
g. 5 4 – 4 1 ——— + ———	**h.** 8 5 – 3 9 ——— + ———

7. Solve. IF you subtract, check the answer by adding (you will not subtract in every problem).

a. Emily picked 29 rows of strawberries and Jim picked 13 rows of strawberries. How many more rows of strawberries did Emily pick?

b. Judith sold 35 tickets for the county fair and Peter sold 62 tickets. How many more tickets did Peter sell than Judith?

c. Judith sold 35 tickets for the county fair and Peter sold 62 tickets. How many tickets did they sell together?

d. A pretty doll with a blue dress costs \$28, and a different doll with a pink dress costs \$12 more than that. How much does that doll cost?

e. Bill bought two bicycle chains for \$18 each and a saddle for \$49. How much was the total cost?

+

Word Problems

"More" in word problems

Study carefully these problems. They all use the word MORE, but they are different! Solve each problem with your teacher or on your own, if you can. In each problem **think** first, "WHO has more?" (If the problems are difficult, drawing illustrations for the situations may help.)

- Anna has 12 sheep. Her neighbor has 7 more sheep than Anna.
  How many sheep does the neighbor have?
- Anna has 7 more sheep than her neighbor. Anna has 12 sheep.
  How many sheep does the neighbor have?
- Anna has 12 sheep, and her neighbor has 7.
  How many more sheep does Anna have than her neighbor?

1. Solve. Think if you need addition or subtraction.

a. Isabella has a flock of 15 goats. Her neighbor Andy has 18 goats.
How many more goats does Andy have than Isabella?

b. Isabella has a flock of 15 goats. Her neighbor Sandy has 8 more goats than Isabella. How many goats does Sandy have?

c. Isabella has 15 goats, which is 5 more than what Henry has.
How many goats does Henry have?

d. Christopher has 27 cows, and Daniel has 16 more than that.
How many cows does Daniel have?

"Fewer" (or "less") in word problems

FEWER is the opposite of MORE. These three problems all use the word FEWER, but they are different! Solve each problem with your teacher or on your own, if you can. In each problem **think** first, "WHO has more?"

- Anna has 12 sheep. Her neighbor has 7 fewer sheep than Anna. How many sheep does the neighbor have?

- Anna has 7 fewer sheep than her neighbor. Anna has 12 sheep. How many sheep does the neighbor have?

- Anna has 7 sheep, and her neighbor has 12. How many fewer sheep does Anna have than her neighbor?

2. Solve. Think if you need addition or subtraction.

a. Joe has 27 tennis balls and Mason has 5 fewer tennis balls than Joe. How many tennis balls does Mason have?

b. Joe has 27 tennis balls, which is 7 less than what Logan has. How many tennis balls does Logan have?

c. Liz wants to buy a blue dress that costs $41.
A white dress, costs $13 less than that.
And a yellow dress costs $3 less than the white dress.

How much does the yellow dress cost?

d. Find how much it is if Liz buys both the blue and the yellow dress.

3. Subtract mentally.

a. 30 – 28 = ________	**d.** 70 – 63 = ________	**g.** 56 – 5 = ________
b. 52 – 30 = ________	**e.** 70 – 6 = ________	**h.** 18 – 7 = ________
c. 98 – 90 = ________	**f.** 100 – 5 = ________	**i.** 46 – 4 = ________

4. Solve. You may need to both add and subtract.

a. Ryan rode his horse for 11 km each day for two days. Zoe rode her horse for 8 km each day for two days.

In two days, how many kilometers less did Zoe ride hers than what Ryan rode his?

b. Mia owns 32 dolls and her friend Ava has less. Actually, Ava has 8 fewer dolls than Mia. How many dolls do the girls have in total?

Puzzle Corner

Find out what number the triangle means. You are solving *equations*!

a. 63 + △ = 71 △ = ________	**b.** 80 − △ = 54 △ = ________	**c.** △ − 10 = 40 △ = ________

Write your own "triangle problems" (equations), and let a friend solve them.

d. △ = ________	**e.** △ = ________	**f.** △ = ________

Mental Subtraction, Part 1

Method 1: Subtract in two parts	
53 − 8 = 53 − 3 − 5 = 50 − 5 = 45	72 − 6 = 72 − 2 − 4 = 70 − 4 = 66
Subtract 8 in two parts: first 3, then 5.	Subtract 6 in two parts: first 2, then 4.

In other words, first subtract to the *previous whole ten*, then the rest.

1. Subtract the elevated number in parts. (First subtract to the previous whole ten; then the rest.)

a. (− 5) 51 − 1 − 4 = ______	**b.** (− 7) 62 − ____ − ____ = ______
c. (− 4) 33 − ____ − ____ = ______	**d.** (− 5) 92 − ____ − ____ = ______
e. (− 6) 75 − ____ − ____ = ______	**f.** (− 7) 63 − ____ − ____ = ______
g. (− 7) 35 − ____ − ____ = ______	**h.** (− 5) 74 − ____ − ____ = ______

2. First subtract the balls that are not in the ten-groups.

a.

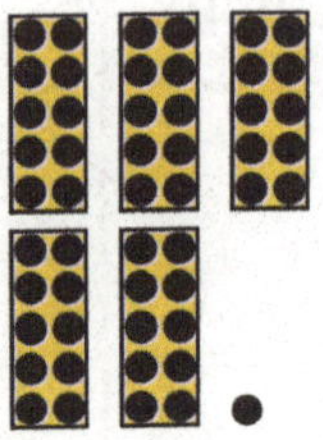

51 − 7 = ______

51 − 5 = ______

51 − 3 = ______

51 − 6 = ______

b.

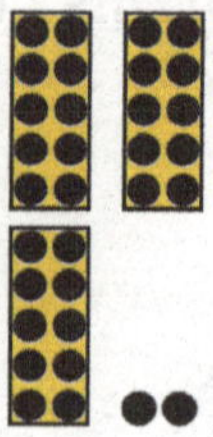

42 − 4 = ______

42 − 5 = ______

42 − 3 = ______

42 − 6 = ______

Method 2: Use known subtraction facts
Since 14 – 6 = 8, we know that the answer to 74 – 6 will end in 8, but it will be sixty-something. So it is 68. Since 15 – 8 = 7, we know that the answer to 55 – 8 will end in 7, but it will be forty-something. So it is 47.

3. Subtract. The first problem in each box is a "helping problem" for the others.

a. 14 – 9 = _______ 24 – 9 = _______ 44 – 9 = _______	**b.** 17 – 8 = _______ 27 – 8 = _______ 37 – 8 = _______	**c.** 12 – 9 = _______ 52 – 9 = _______ 32 – 9 = _______
d. 15 – 9 = _______ 65 – 9 = _______ 45 – 9 = _______	**e.** 13 – 8 = _______ 33 – 8 = _______ 93 – 8 = _______	**f.** 16 – 8 = _______ 86 – 8 = _______ 36 – 8 = _______

4. **a.** Amy has $32. She bought a comic book for $7.
How much does she have now?

b. Peter had $29. A toy train he wants costs $39.
Mom paid him $5 for working. How much more
does Peter now need to buy the train?

c. A flower shop has 55 roses. Eight of them are white,
and the rest are red. How many are red?

5. Use either method from this lesson to subtract.

a.	**b.**	**c.**	**d.**
34 – 5 = ______	65 – 9 = ______	51 – 8 = ______	62 – 7 = ______
73 – 7 = ______	36 – 8 = ______	93 – 6 = ______	83 – 8 = ______

Mental Subtraction, Part 2

Method 3: Add.	
You can "add backwards". This works well if the two numbers are close to each other. Instead of subtracting, think how much you need to add to the number being subtracted (the subtrahend) in order to get the number you are subtracting from (the minuend).	
Think: 84 + ____ = 92 (84 and how many more makes 92?) + __ __ 92 – 84 = _____	Think: 25 + ____ = 75 (25 and how many more makes 75?) + ____ 75 – 25 = _____

1. To find these differences, think of adding more.

+ 8 **a.** 92 – 84 = ______ (Think: 84 + ____ = 92)	+ ____ **b.** 51 – 49 = ______ (Think: 49 + ___ = 51)	+ ____ **c.** 76 – 69 = ______ (Think: 69 + ___ = 76)
d. 32 – 28 = ______ **e.** 22 – 14 = ______ **f.** 53 – 46 = ______	**g.** 90 – 83 = ______ **h.** 64 – 56 = ______ **i.** 72 – 65 = ______	**j.** 100 – 95 = ______ **k.** 64 – 55 = ______ **l.** 44 – 37 = ______

2. Solve. Think if you need addition or subtraction—or both.

a. Jerry has 46 toy cars. Larry has 7 more than Jerry, and Mickey has 7 less than Jerry. How many toy cars does Larry have?

And Mickey?

b. Andy has $33. He bought a gift for his mom that cost $28. Then, Andy got $5 from his dad for helping with car repairs. How much money does Andy have now?

Method 4: Subtract in parts: tens and ones

Break the number being subtracted into its tens and ones. Subtract in parts.

53 − 21 = 53 − 20 − 1 = 33 − 1 = 32 First subtract 20, then 1.	87 − 46 = 87 − 40 − 6 = 47 − 6 = 41 First subtract 40, then 6.

3. Solve.

a. 78 − 22 THINK: 78 − 20 − 2 = ______	**b.** 56 − 31 THINK: 56 − 30 − 1 = ______	**c.** 46 − 25 46 − ______ − ____ = ______
d. 66 − 43 66 − ______ − ____ = ______	**e.** 28 − 12 28 − ______ − ____ = ______	**f.** 84 − 52 84 − ______ − ____ = ______
g. 99 − 66 = ______	**h.** 47 − 34 = ______	**i.** 78 − 67 = ______

4. **a.** Noah counted books on his bookshelf. One shelf had 34 books. Another had 42. How many more books did that shelf have than the first?

b. Noah took four books from the second shelf and put them on the first one. Now how many more books does the second shelf have than the first?

5. Draw arrows to illustrate the subtraction on the number line.

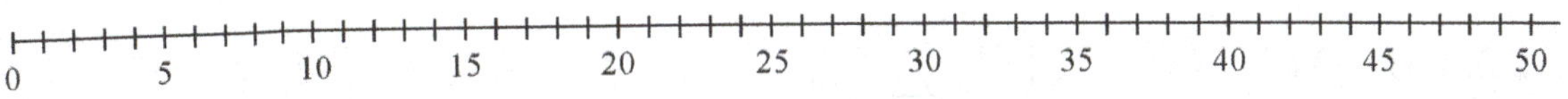

35 − 18 = ______

<u>Method 5:</u> **Subtract in parts: tens and ones**	
Break BOTH the number you subtract from AND the number being subtracted into its tens and ones. Subtract the tens. Subtract the ones.	
53 − 21 = 50 − 20 and 3 − 1 = 30 and 2 = 32	76 − 33 = 70 − 30 and 6 − 3 = 40 and 3 = 43

6. Solve.

a. 67 − 53 60 − 50 and 7 − 3 = ______	**b.** 92 − 31 90 − 30 and 2 − 1 = ______	**c.** 85 − 22 80 − 20 and ___ − ___ = ______
d. 88 − 62 = ______	**e.** 57 − 23 = ______	**f.** 79 − 17 = ______

7. **a.** Terry is on page 52 of her book. The book has a total of 95 pages. How many pages does she have left to read?

b. Terry reads nine pages more. Now how many pages does she have left to read?

8. Devise your own method for these subtractions. Explain how your method works.

a. 52 − 36	**b.** 81 − 47

Puzzle Corner The triangle and square represent "mystery numbers." Find them. Guess and check! Then change your guess.

a. △ + △ + 10 = 34 △ = ______	**b.** □ + △ = 22 □ − △ = 4 △ = ______, □ = ______	**c.** □ + △ = 22 □ + □ = 36 △ = ______, □ = ______

Regrouping One Ten as Ten Ones with 3-Digit Numbers

You have already learned how to regroup one ten as ten ones (borrow a ten) with two-digit numbers. It happens the same way with three-digit numbers.

We need to subtract 229.... but we can't take away nine ones because there are only five ones. →

"Break down" one ten as 10 ones. Now we can subtract! Take away 2 hundreds, 2 tens, and 9 ones.

What is left? _________

→

365 → **300 + 50 + 15**

1. Break down ONE ten into ten ones (regroup). Draw squares for hundreds, sticks for tens, and dots for ones. Then take away (subtract) what is asked.

a.

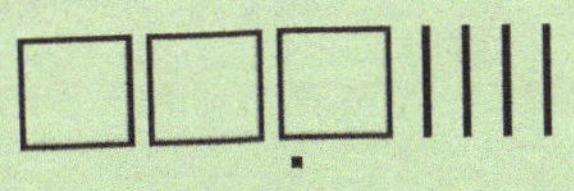

→

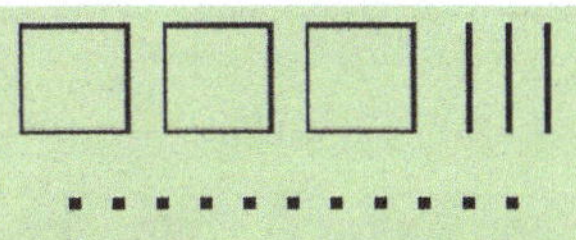

341 → 300 + 30 + 11

Take away 127. What is left? __________

b.

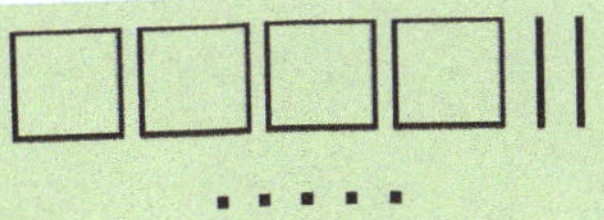

→

425 → 400 + ______ + ____

Take away 218. What is left? __________

c.

→

267 → 200 + ______ + ____

Take away 139. What is left? __________

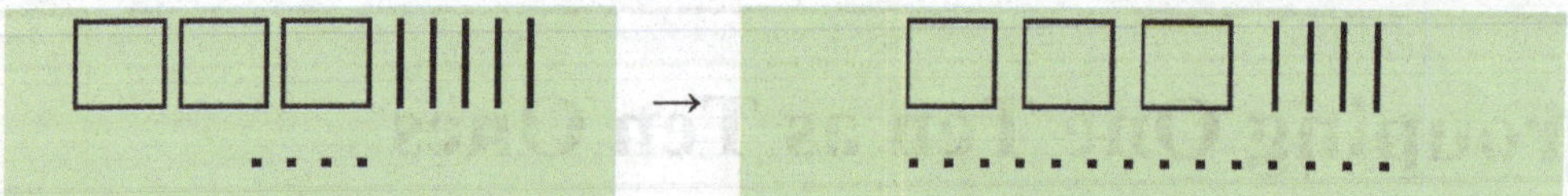

354 → 300 + 40 + 14

Take away 136. What is left?

h	t	o
	4	14
3	~~5~~	4
− 1	3	6
2	1	8

COMPARE the regrouping process in the pictures and how it is written in columns. It happens the same way as with two-digit numbers: Instead of 5 tens and 4 ones, we write 4 tens and 14 ones. Then we can subtract in the ones column.

2. Subtract. Regroup in the tens.

a.	b.	c.
8 6 1 − 4 6	5 9 2 − 1 2 5	7 7 3 − 3 5 6

3. Subtract. Regroup if necessary. Check each subtraction by *adding your answer and the number you subtracted.*

a. 1 13 8 ~~2~~ ~~3~~ − 2 0 5 6 1 8 6 1 8 + 2 0 5	**b.** 6 7 1 − 4 5 5 + 4 5 5
c. 4 2 0 − 1 1 5 +	**d.** 5 1 6 − 2 0 8 +
e. 7 3 3 − 6 2 7 +	**f.** 6 8 7 − 3 0 9 +

4. Continue the patterns.

a.	b.
120 + 120 = 240	1 + 119 = 120
121 + ______ = 240	2 + ______ = 120
122 + ______ = 240	3 + ______ = 120
______ + ______ = 240	______ + ______ = 120
______ + ______ = 240	______ + ______ = 120
______ + ______ = 240	______ + ______ = 120
______ + ______ = 240	______ + ______ = 120
______ + ______ = 240	______ + ______ = 120
______ + ______ = 240	______ + ______ = 120
THINK: How long can you continue this pattern?	THINK: How long can you continue this pattern?

5. Solve the problems.

a.	b.	c.
There are 365 days in one year. How many days are there in two years?	Jack's family is driving from Easttown to Middletown, (a distance of 173 km). They have driven 69 km. How far do they still have to go?	Mike is on page 235 of his book. The book has 581 pages. How many pages does he still have to read?

Regrouping One Hundred as 10 Tens

We need to subtract 170.... but we cannot take away seven tens because there are only two tens.	→	"Break down" one HUNDRED as 10 tens. Now we can subtract! Take away 1 hundred and 7 tens. What is left? ________
	→	
320	→	**2 hundreds + 12 tens**

1. Break down one hundred into 10 tens (regroup). Draw squares for hundreds, sticks for tens, and dots for ones. Then take away (subtract) what is asked.

a.  → 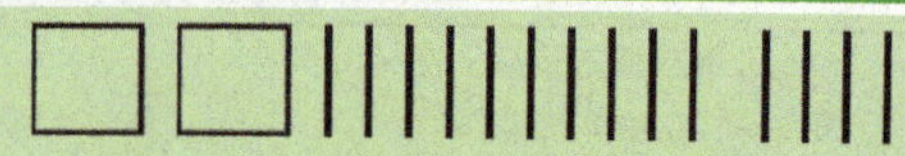

340 → _____ hundreds + _____ tens

Take away 180. What is left? _________

b.  →

410 → _____ hundreds + _____ tens

Take away 250. What is left? _________

c. 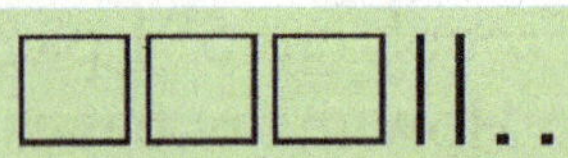 →

322 → _____ hundreds + _____ tens + ______ ones

Take away 171. What is left? _________

d.  →

254 → _____ hundreds + _____ tens + ______ ones

Take away 174. What is left? _________

2. First, regroup 1 hundred as ten tens. Then subtract.

a. 4 hundreds 5 tens 7 ones ⇨

	3	15 tens	7 ones
–	2 hundreds	8 tens	2 ones
	1	7 tens	5 ones

b. 7 hundreds 2 tens 1 one ⇨

	___ hundreds	___ tens	___ one
–	3 hundreds	6 tens	1 one
	___ hundred	___ tens	___ ones

c. 3 hundreds 2 tens 0 ones ⇨

	___ hundreds	___ tens	___ ones
–	2 hundreds	5 tens	0 ones
	___ hundred	___ tens	___ ones

d. 7 hundreds 0 tens 6 ones ⇨

	___ hundreds	___ tens	___ ones
–	6 hundreds	2 tens	2 ones
	___ hundred	___ tens	___ ones

e. 8 hundreds 0 tens 3 ones ⇨

	___ hundreds	___ tens	___ ones
–	5 hundreds	3 tens	1 one
	___ hundred	___ tens	___ ones

3. How to regroup when subtracting 947 – 282 (below)? Fill in Jill's explanation.

It would be easy, except I cannot subtract ____ tens from ____ tens. So, I need to take one of the ____ hundreds and break it down as tens. So, now I will have only ____ hundreds but I will now get ___ tens. Now I can subtract.

9 hundreds 4 tens 7 ones ⇨

	___ hundreds	___ tens	___ ones
–	2 hundreds	8 tens	2 ones
	___ hundred	___ tens	___ ones

Compare how we write the regrouping when subtracting in columns.	
5 hundreds 4 tens 7 ones ⇨ 4 hundreds 14 tens 7 ones − 1 hundred 5 tens 2 ones 3 hundreds 9 tens 5 ones	4 14 ~~5~~ 4 7 − 1 5 2 3 9 5

4. Fill in. Subtract both ways.

a. 4 hundreds 5 tens 6 ones ⇨ ☐ hundreds ☐ tens ☐ ones − 2 hundreds 7 tens 2 ones ☐ hundreds ☐ tens ☐ ones	4 5 6 − 2 7 2
b. 6 hundreds 0 tens 5 ones ⇨ ☐ hundreds ☐ tens ☐ ones − 4 hundreds 3 tens 3 ones ☐ hundreds ☐ tens ☐ ones	6 0 5 − 4 3 3

5. Subtract.

a. 9 2 6 − 1 4 6	**b.** 5 2 9 − 9 5	**c.** 4 1 4 − 3 2 2	**d.** 7 7 3 − 5 3 6
e. 6 7 0 − 2 2 6	**f.** 7 0 8 − 1 5 6	**g.** 5 0 3 − 3 4 1	**h.** 7 4 8 − 3 7 6

6. Solve the problems.

a. Max has two books to read. The first book has 270 pages, and the second book has 60 fewer pages than the first. How many pages does the second book have?

b. Liz and Hannah played a game. Hannah got 192 points and Liz got 433 points. How many more points did Liz get than Hannah?

c. Again, Liz and Hannah played a game. This time Liz got 215 points and Hannah got 93 points more than Liz. So, how many points did Hannah get?

d. Denny and Micah dug up some worms for bait before they went fishing. Denny got 14 worms, which was 11 fewer worms than what Micah got. How many worms did Micah get?

What was the total number of worms that both boys got?

Puzzle Corner Figure out the missing numbers in these subtractions! You might need to regroup.

$$\begin{array}{r} \square\ \square\ 5 \\ -\ 1\ \ 5\ \square \\ \hline 2\ \ 9\ \ 2 \end{array} \qquad \begin{array}{r} 6\ \square\ 4 \\ -\ \square\ 5\ \square \\ \hline 3\ \ 2\ \ 6 \end{array} \qquad \begin{array}{r} 9\ \square\ \square \\ -\ \square\ 5\ \ 5 \\ \hline 7\ \ 2\ \ 6 \end{array} \qquad \begin{array}{r} 9\ \ 6\ \square \\ -\ \square\ 5\ \ 5 \\ \hline 5\ \square\ 5 \end{array}$$

Graphs and Problems

1. The table lists the eye colors of some children. Draw the bars for the bar graph.

9	4	10	16	11
blue	green	gray	brown	hazel

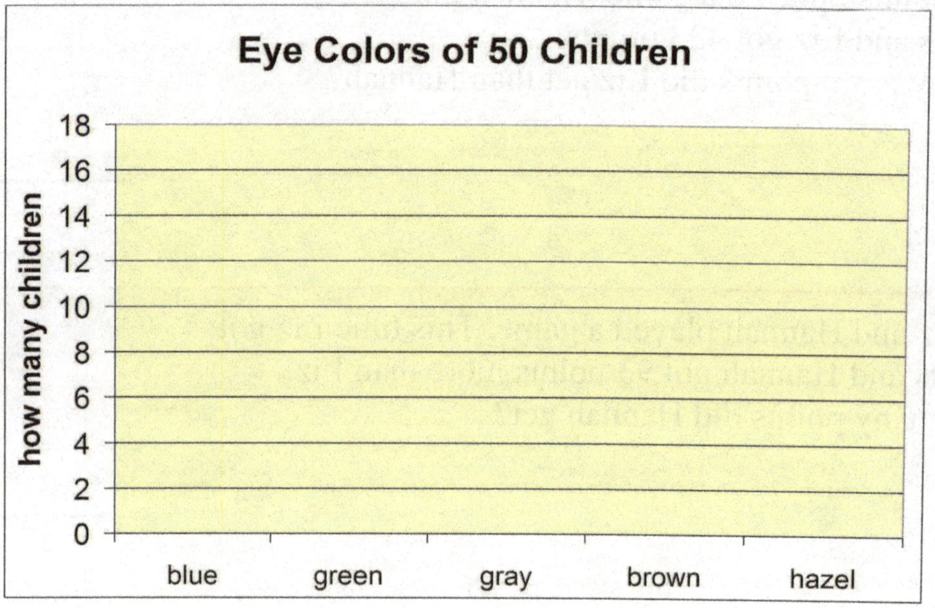

a. How many children have either brown or hazel eyes?

b. How many more children have brown eyes than have green eyes?

c. How many children do not have blue eyes?

2. Solve the word problems. Add or subtract in columns if you need to.

a. Jim has 62 marbles, Peter has 28, and Ed has 33 marbles. Peter and Ed put their marbles together. Now do they have more than Jim?

b. You were on page 48 in a book that has a total of 95 pages. Then you read 10 more pages. How many pages have you read now?

How many pages do you have left to read?

3. The bar graph below tells us the counts for different kinds of animals in a zoo. Answer the questions. Use the grids for additions and subtractions if you need to.

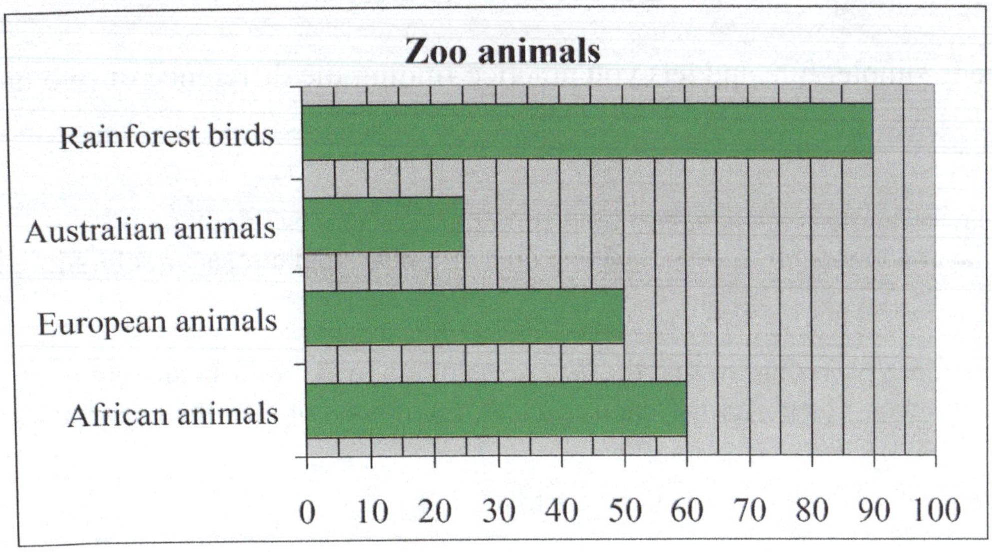

a. How many rainforest birds does the zoo have?

b. How many Australian animals are there?

c. How many Australian and African animals does the zoo have together?

d. How many more African animals does the zoo have than Australian animals?

e. How many Australian, European, and African animals are there?

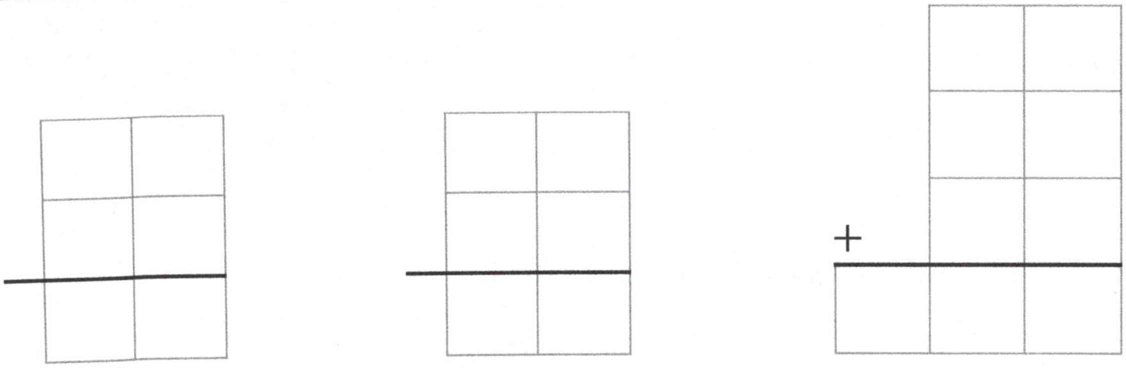

4. Break the number you are subtracting into tens and ones, and subtract in parts.

a. 45 – 23 / \\ 45 – 20 – 3 = ______	**b.** 76 – 15 / \\ 76 – 10 – 5 = ______	**c.** 47 – 15

Euclid's Game

Euclid's game is simple, fun, and lets you practice finding the difference of two numbers!

Rules:

1. The first player chooses *any number* on the 100-chart and circles or colors it.
2. The next player chooses *any other number* on the 100-chart and circles or colors it.

After this, the numbers get marked by crossing them out.

During his turn, each player has to find the *difference* of any two numbers already marked, and mark that number. The player can choose *any* two numbers for this; they just have to be already marked numbers.

The player who cannot find any more numbers to mark is the loser.

Use the 100-chart on the next page as a game board. You can print it again for every new game. Alternatively, you may write the 100-chart on paper, of course.

Example. Initially Jane chooses 28 and Joe chooses 9. After that:

Jane: I mark 19 since it is the difference of 28 and 9.
Joe: I mark 10 since it is the difference of 19 and 9.
Jane: I mark 18 since it is the difference of 28 and 10.
Joe: I mark 1 since it is the difference of 19 and 18.
... and so on.

You may continue Jane's and Joe's play if you wish.

Eventually you should see all of the numbers from 1 to 28 marked, with Jane as the loser.

~~1~~	2	3	4	5	6	7	8	9	~~10~~
11	12	13	14	15	16	17	~~18~~	~~19~~	20
21	22	23	24	25	26	27	28	29	30
31	32	33	34	35	36	37	38	39	40
41	42	43	44	45	46	47	48	49	50
51	52	53	54	55	56	57	58	59	60
61	62	63	64	65	66	67	68	69	70
71	72	73	74	75	76	77	78	79	80
81	82	83	84	85	86	87	88	89	90
91	92	93	94	95	96	97	98	99	100

Questions to ponder after you have played a few games:

1. Suppose that 28 and 9 are chosen as the initial numbers, like in Jane's and Joe's game. Can Jane and Joe ever mark a number that is more than 28?
2. Let's say that the two initial numbers are both even. What can you say about the numbers that get marked in the game?
3. Suppose that the two initial numbers are both multiples of 5, such as 55 and 30. What can you say about the numbers that get marked in the game?
4. Can you mark off *all* of the the numbers on the 100-chart during the game, if the initial numbers are (you can try these out): **a.** 90 and 7? **b.** 100 and 1? **c.** 100 and 10? **d.** 100 and 13?

100-Charts for Euclid's game or other uses

1	2	3	4	5	6	7	8	9	10
11	12	13	14	15	16	17	18	19	20
21	22	23	24	25	26	27	28	29	30
31	32	33	34	35	36	37	38	39	40
41	42	43	44	45	46	47	48	49	50
51	52	53	54	55	56	57	58	59	60
61	62	63	64	65	66	67	68	69	70
71	72	73	74	75	76	77	78	79	80
81	82	83	84	85	86	87	88	89	90
91	92	93	94	95	96	97	98	99	100

1	2	3	4	5	6	7	8	9	10
11	12	13	14	15	16	17	18	19	20
21	22	23	24	25	26	27	28	29	30
31	32	33	34	35	36	37	38	39	40
41	42	43	44	45	46	47	48	49	50
51	52	53	54	55	56	57	58	59	60
61	62	63	64	65	66	67	68	69	70
71	72	73	74	75	76	77	78	79	80
81	82	83	84	85	86	87	88	89	90
91	92	93	94	95	96	97	98	99	100

1	2	3	4	5	6	7	8	9	10
11	12	13	14	15	16	17	18	19	20
21	22	23	24	25	26	27	28	29	30
31	32	33	34	35	36	37	38	39	40
41	42	43	44	45	46	47	48	49	50
51	52	53	54	55	56	57	58	59	60
61	62	63	64	65	66	67	68	69	70
71	72	73	74	75	76	77	78	79	80
81	82	83	84	85	86	87	88	89	90
91	92	93	94	95	96	97	98	99	100

1	2	3	4	5	6	7	8	9	10
11	12	13	14	15	16	17	18	19	20
21	22	23	24	25	26	27	28	29	30
31	32	33	34	35	36	37	38	39	40
41	42	43	44	45	46	47	48	49	50
51	52	53	54	55	56	57	58	59	60
61	62	63	64	65	66	67	68	69	70
71	72	73	74	75	76	77	78	79	80
81	82	83	84	85	86	87	88	89	90
91	92	93	94	95	96	97	98	99	100

1	2	3	4	5	6	7	8	9	10
11	12	13	14	15	16	17	18	19	20
21	22	23	24	25	26	27	28	29	30
31	32	33	34	35	36	37	38	39	40
41	42	43	44	45	46	47	48	49	50
51	52	53	54	55	56	57	58	59	60
61	62	63	64	65	66	67	68	69	70
71	72	73	74	75	76	77	78	79	80
81	82	83	84	85	86	87	88	89	90
91	92	93	94	95	96	97	98	99	100

1	2	3	4	5	6	7	8	9	10
11	12	13	14	15	16	17	18	19	20
21	22	23	24	25	26	27	28	29	30
31	32	33	34	35	36	37	38	39	40
41	42	43	44	45	46	47	48	49	50
51	52	53	54	55	56	57	58	59	60
61	62	63	64	65	66	67	68	69	70
71	72	73	74	75	76	77	78	79	80
81	82	83	84	85	86	87	88	89	90
91	92	93	94	95	96	97	98	99	100

Review

1. Add.

a.
$$\begin{array}{r} 2\ 1\ 5 \\ +\ 4\ 7\ 7 \\ \hline \end{array}$$

b.
$$\begin{array}{r} 1\ 9\ 2 \\ +\ 2\ 2\ 5 \\ \hline \end{array}$$

c.
$$\begin{array}{r} 3\ 0\ 3 \\ 1\ 2\ 8 \\ +\ 2\ 8\ 7 \\ \hline \end{array}$$

d.
$$\begin{array}{r} 4\ 0\ 9 \\ 2\ 1\ 9 \\ +\ 1\ 3\ 6 \\ \hline \end{array}$$

2. Sarah bought three bicycles for her children.
Each bicycle cost $154.
How much was the total cost?

+

3. Add mentally. THINK of the new hundred you might get from adding the tens.

a.	b.	c.
80 + 40 = ________	90 + 90 = ________	690 + 50 = ________
780 + 40 = ________	240 + 50 = ________	470 + 80 = ________

4. Find how many feet it is if you walk all of the way around this rectangle.

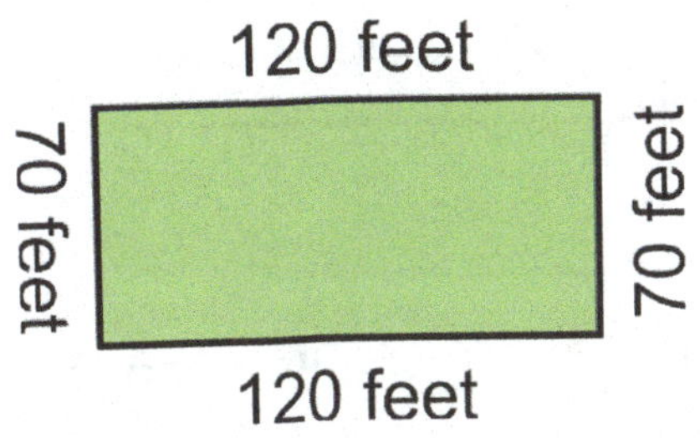

+

5. Subtract. Regroup if necessary. Check each subtraction by *adding your answer and the number you subtracted.*

a. $88 - 54 = \_\_\_$ Check: $\_\_\_ + 54 = \_\_\_$	**b.** $63 - 48 = \_\_\_$ Check: $\_\_\_ + \_\_\_ = \_\_\_$
c. $84 - 49 = \_\_\_$ Check: $\_\_\_ + \_\_\_ = \_\_\_$	**d.** $882 - 159 = \_\_\_$ Check: $\_\_\_ + \_\_\_ = \_\_\_$
e. $556 - 391 = \_\_\_$ Check: $\_\_\_ + \_\_\_ = \_\_\_$	**f.** $550 - 246 = \_\_\_$ Check: $\_\_\_ + \_\_\_ = \_\_\_$

6. Subtract using mental math methods.

a. 15 − 7 = _____ 55 − 7 = _____	**b.** 13 − 5 = _____ 93 − 5 = _____	**c.** 82 − 77 = _____ 45 − 41 = _____
d. 80 − 71 = _____ 100 − 95 = _____	**e.** 56 − 40 = _____ 56 − 43 = _____	**f.** 78 − 35 = _____ 33 − 4 = _____

7. Find what numbers are missing.

a. $2\_4 + 477 = 731$

b. $5\_9 + \_25 = 914$

c. $20\_ + 6\_6 = 892$

d. $68\_ + \_19 = 900$

8. Solve.

a. Some people are riding on the bus. At the bus stop, 13 people get on. Now there are 52 people on the bus. How many were there originally?

b. Molly has 23 stuffed toys that she likes, and 16 that she does not like.

How many stuffed toys does Molly have?

c. Molly gave the 16 toys she does not like to her sister Annie. Now, Annie has 33 toys.

How many toys did Annie have before?

d. Jessica had 465 points in a computer game. She played and got 145 more points. Then she also got a 90-point bonus! How many points does Jessica have now?

+

e. Olivia did 26 jumping jacks, which was 14 fewer jumping jacks than what her brother Aaron did. How many jumping jacks did Aaron do?

9. **a.** Fill in the table with how many points the children got in the game.

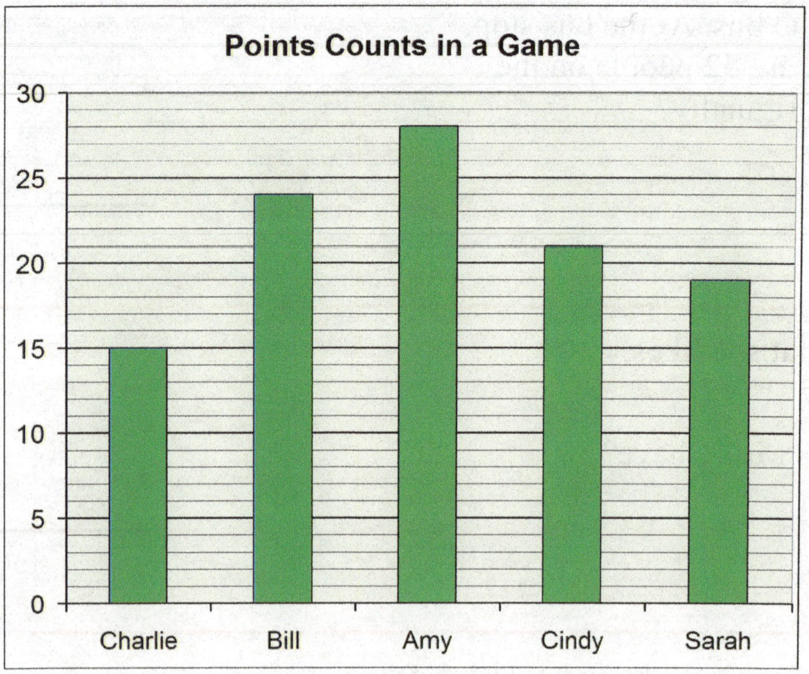

CHILD	POINTS
Charlie	15
Bill	
Amy	
Cindy	
Sarah	

b. How many fewer points did Bill get than Amy?

c. How many more points did Cindy get than Charlie?

Puzzle Corner

Can you place numbers from 1 through 12 into the circles so that the sum of each connecting line is 26?

Hint: The numbers that go in the top corners are 7 and 6, and the numbers that go in the bottom corners are 5 and 8.

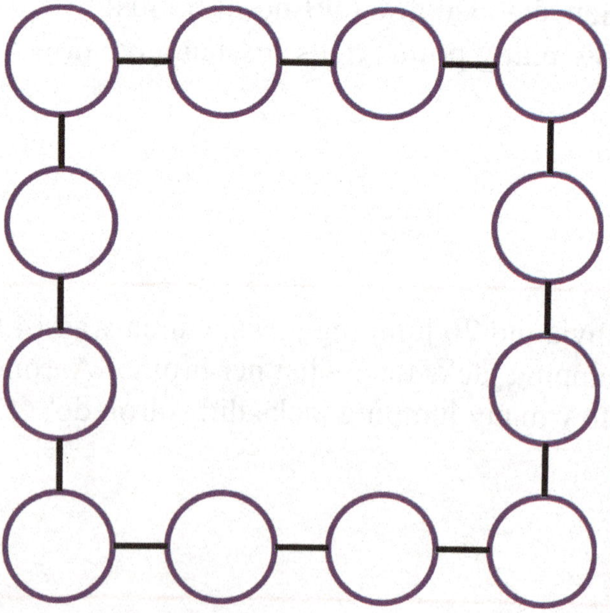

Answers

Adding 3-Digit Numbers in Columns, p. 9

1. a. 308 + 132 = 440　　b. 408 + 233 = 641
   c. 226 + 237 = 463　　d. 308 + 157 = 465

2. a. 811　b. 763　c. 772　d. 583
   c. 866　f. 989　g. 933　h. 965

3. a. Now there are 675 workers. 547 + 128 = 675
   b. There are 580 balloons left in the bags. 600 – 20 = 580
   c. The total cost is $348. 116 + 116 + 116 = 348

Puzzle Corner:

6 3 1	2 2 5	1 1 6	2 4 7
+ 1 4 5	+ 3 5 5	+ 4 5 7	+ 7 4 5
7 7 6	5 8 0	5 7 3	9 9 2

Regrouping 10 Tens as a Hundred, p. 11

1. Check the student's work as far as circling the tens.
   a. 130　b. 310
   c. 530　d. 355
   e. 741　f. 656

2. a. 130
   b. 320
   c. 413
   d. 646
   e. 715

3. a. 110, 210, 310　b. 110, 210, 310　c. 140, 340, 540

4. a. 110　b. 310　c. 117　d. 444　e. 815
   f. 139　g. 417　h. 856　i. 729　j. 869

5. a. 110, 170, 210　b. 140, 330, 180　c. 340, 270, 210

6.

a.	b.	c.	d.	e.
167	240	391	653	375
+ 192	+ 182	+ 422	+ 193	+ 184
359	422	813	846	559

7. When the elephant and the giraffe ran a race, who won? The elephant, because the giraffe was in the refrigerator. (This is a continuation to a silly joke question, which goes: "How do you get a giraffe into a refrigerator?" — Open the door, put the giraffe in, close the door.)

Add in Columns: Regrouping Twice, p. 15

1. a. 58 + 83 = 141　b. 86 + 79 = 165
   c. 155 + 69 = 224　d. 177 + 136 = 313

2. a. 241　b. 342　c. 422　d. 652

3. Mary circled 10 tens to make another hundred, and she circled the ones to form a 10. Then she added the tens with the new group, which made 40, and she added the hundreds, which made 600. This gave her a total of 640.

Add in Columns: Regrouping Twice, cont.

4. a. 767 b. 524 c. 810 d. 651 e. 860 f. 601
g. 624 h. 819 i. 759 j. 931 k. 832 l. 930

5. *First I check the hundreds: 600 + 300 makes 900.*
Then I add the tens, and I get 50 + 30 = 80 .
Lastly in the ones, there's just 0 and 1, which is 1.
Okay, so I have these parts: 900, 80, and 1, so that makes 981.

6. a. 148 + 148 = 296 miles.
b. \$128 + \$129 + \$549 = \$896 was the total cost.
c. 156 + 156 + 124 = 436 feet.

Puzzle corner:

$$\begin{array}{r} \boxed{2}\ 3\ \boxed{2} \\ +\ 1\ \boxed{5}\ 9 \\ \hline 3\ 9\ 1 \end{array} \qquad \begin{array}{r} 2\ \boxed{8}\ \boxed{1} \\ +\ \boxed{2}\ 3\ 6 \\ \hline 5\ 1\ 7 \end{array} \qquad \begin{array}{r} 1\ 6\ 9 \\ +\ \boxed{5}\ 5\ \boxed{5} \\ \hline 7\ 2\ 4 \end{array} \qquad \begin{array}{r} \boxed{1}\ 8\ 8 \\ +\ 7\ \boxed{1}\ \boxed{2} \\ \hline 9\ 0\ 0 \end{array}$$

Regrouping in Subtraction, Part 1, p. 19

1.

a. 3 tens 0 ones ⇨ 2 tens 10 ones	b. 3 tens 6 ones ⇨ 2 tens 16 ones
c. 5 tens 1 one ⇨ 4 tens 11 ones	d. 5 tens 5 ones ⇨ 4 tens 15 ones
e. 4 tens 0 ones ⇨ 3 tens 10 ones	f. 3 tens 7 ones ⇨ 2 tens 17 ones

2. a. What is left? 2 tens 8 ones
b. 5 tens 4 ones → 4 tens 14 ones. What is left? 2 tens 7 ones.
c. 4 tens 3 ones → 3 tens 13 ones. What is left? 1 ten 8 ones.
d. 6 tens 1 one → 5 tens 11 ones. What is left? 1 ten 7 ones.

3.
a. 5 tens 5 ones ⇨ 4 tens 15 ones − 3 tens 7 ones = 1 ten 8 ones
b. 7 tens 2 ones ⇨ 6 tens 12 ones − 3 tens 5 ones = 3 tens 7 ones
c. 6 tens 0 ones ⇨ 5 tens 10 ones − 2 tens 7 ones = 3 tens 3 ones
d. 6 tens 4 ones ⇨ 5 tens 14 ones − 3 tens 8 ones = 2 tens 6 ones
e. 7 tens 6 ones ⇨ 6 tens 16 ones − 4 tens 7 ones = 2 tens 9 ones
f. 5 tens 0 ones ⇨ 4 tens 10 ones − 2 tens 2 ones = 2 tens 8 ones
g. 8 tens 1 one ⇨ 7 tens 11 ones − 6 tens 5 ones = 1 ten 6 ones
h. 6 tens 3 ones ⇨ 5 tens 13 ones − 2 tens 8 ones = 3 tens 5 ones

4. a. Jessica now has 13 pencils. 27 – 10 – 4 = 13.
b. Three more pencils. 13 – 10 = 3.
c. Nine more pencils. 13 – 4 = 9.

Regrouping in Subtraction, Part 2, p. 22

1.

a. 6 tens 0 ones → 5 tens 10 ones

Take away
3 tens, 9 ones.

	6	0
−	3	9
	2	1

b. 7 tens 1 one → 6 tens 11 ones

Take away
1 ten, 6 ones.

	7	1
−	1	6
	5	5

c. 3 tens, 5 ones → 2 tens 15 ones

Take away
1 ten, 7 ones.

	3	5
−	1	7
	1	8

d. 8 tens → 7 tens 10 ones

Take away
3 tens, 4 ones.

	8	0
−	3	4
	4	6

e. 7 tens, 6 ones → 6 tens 16 ones

Take away
4 tens, 8 ones.

	7	6
−	4	8
	2	8

f. 9 tens → 8 tens 10 ones

Take away
5 tens, 1 one.

	9	0
−	5	1
	3	9

g. 5 tens, 4 ones → 4 tens 14 ones

Take away
2 tens, 5 ones.

	5	4
−	2	5
	2	9

h. 8 tens → 7 tens 10 ones

Take away
4 tens, 7 ones.

	8	0
−	4	7
	3	3

i. 7 tens, 4 ones → 6 tens 14 ones

Take away
3 tens, 8 ones.

	7	4
−	3	8
	3	6

j. 4 tens 7 ones → 3 tens 17 ones

Take away
2 tens, 9 ones.

	4	7
−	2	9
	1	8

2. In this exercise, the child is asked to check each subtraction with addition. Adding back is just one method of checking. Estimating the result beforehand is another method of checking - it won't tell you if you got it exactly right but it does check that the result is at least reasonable. Yet another way is to solve the same problem with some different method (there might not always be a "different method"). It is not good that children get used to always checking answers for math problems from the back of the book or from their teacher. They should get used to checking the results themselves. You can encourage the students to check their results even if there is no written instruction to do so.

a. 29 b. 62 c. 27 d. 55 e. 57 f. 26 g. 38 h. 39 i. 22 j. 38 k. 13 l. 38

Puzzle corner:

	9	3
−	1	**8**
	7	5

	8	**3**
−	**6**	7
	1	6

	8	0
−	3	**8**
	4	2

	8	**2**
−	1	4
	6	8

	6	2
−		**9**
	5	3

Regrouping in Subtraction, Part 3, p. 25

1. a. no b. yes c. yes

2. a. YES; 44 b. NO; 25 c. YES; 26 d. YES; 52 e. NO; 72 f. YES; 24 g. NO; 10 h. YES; 19

3. a. 44, 48 b. 6, 5 c. 92, 4

4. a. 45 – 27 = 18 b. 34 – 13 = 21 c. 31 – 9 – 8 = 14

5. a. 22 – 9 = 13

b. 36 – 12 = 24

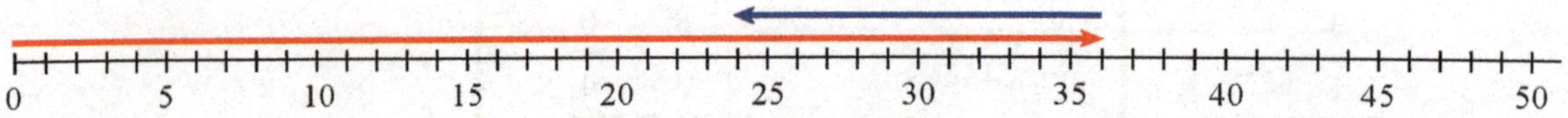

c. 44 – 17 = 27

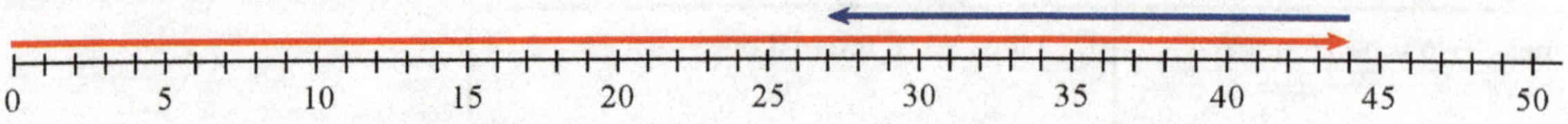

6. a. 94 b. 57, 82 c. 12, 61 d. 42, 99 e. 37, 60 f. 18, 66 g. 13, 54 h. 46, 85

7. a. Emily picked 16 rows more.
   b. Peter sold 27 more tickets.
   c. They sold 97 tickets.
   d. The doll with the pink dress costs $40.
   e. The total cost was $85.

Word Problems, p. 29

1. a. Andy has three more goats than Isabella. b. Sandy has 23 goats.
   c. Henry has ten goats. d. Daniel has 43 cows.

2. a. Mason has 22 tennis balls.
   b. Logan has 34 tennis balls.
   c. The yellow dress costs $25.
   d. Together, the blue and yellow dresses would cost $66.

3. a. 2 b. 22 c. 8 d. 7 e. 64 f. 95 g. 51 h. 11 j. 42

4. a. 11 – 8 = 3 + 3 = 6 km Zoe rode 6 km less than Ryan for 2 days.
   b. 32 – 8 = 24 + 32 = 56 Together, the two girls have 56 dolls.

Puzzle corner: a.  = 8 b.  = 26 c.  = 50 d.-f. Answers will vary.

Mental Subtraction, Part 1, p. 32

1. a. (51 – 1) – 4 = 46 b. (62 – 2) – 5 = 55 c. (33 – 3) – 1 = 29 d. (92 – 2) – 3 = 87
   e. (75 – 5) – 1 = 69 f. (63 – 3) – 4 = 56 g. (35 – 5) – 2 = 28 h. (74 – 4) – 1 = 69

2. a. 44, 46, 48, 45 b. 38, 37, 39, 36

3. a. 5, 15, 35 b. 9, 19, 29 c. 3, 43, 23 d. 6, 56, 36 e. 5, 25, 85 f. 8, 78, 28

4. a. Amy has $25 now.
   b. Peter needs $5 more.
   c. 47 are red.

5. a. 29, 66 b. 56, 28 c. 43, 87 d. 55, 75

Mental Subtraction, Part 2, p. 34

1. a. 8 b. 2 c. 7 d. 4 e. 8 f. 7
   g. 7 h. 8 i. 7 j. 5 k. 9 l. 7

2. a. Larry has 53 cars. 46 + 7 = 53. Mickey has 39 cars. 46 – 7 = 39.
   b. Andy has $10 now. After he bought the gift, he had $33 – $28 = $5. Then he got $5 more so he had $5 + $5 = $10.

3. a. 56 b. 25 c. 21 d. 23 e. 16 f. 32 g. 33 h. 13 i. 11

4. a. There were 8 more books. 42 – 34 = 8.
   b. They both now have 38 books. The second shelf has now 42 – 4 = 38, and the first shelf has 34 + 4 = 38.

5. 35 – 18 = 17

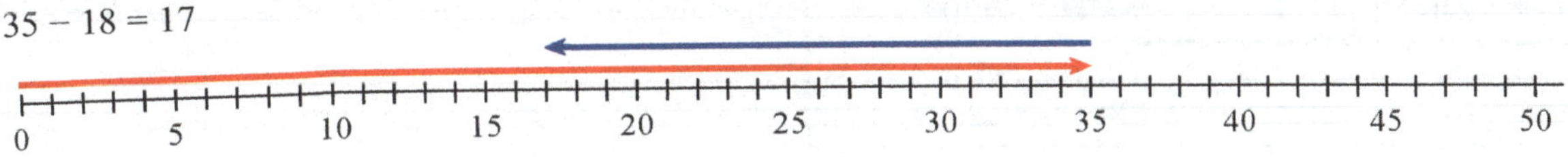

6. a. 14 b. 61 c. 63 d. 26 e. 34 f. 62

7. a. Terry has 43 pages left to read. 95 – 52 = 43.
   b. Now she has 34 pages left to read. You can solve this in many ways. For example: she has now read 52 + 9 = 61 pages. So, she has 95 – 61 = 34 pages left. Or, since she had 43 pages left to read earlier, now she has 43 – 9 = 34 pages left.

8. There are various methods you can use, such as breaking a ten into ones to subtract; or adding up to the next number; or subtracting in parts. If the student cannot answer this question, then he/she needs to review what has already been taught in previous lessons.

   a. 16 b. 34

Puzzle corner:

a. △ = 12 b. △ = 9 □ = 13 c. △ = 4 □ = 18

Regrouping One Ten as Ten Ones with 3-Digit Numbers, p. 37

1.

a.
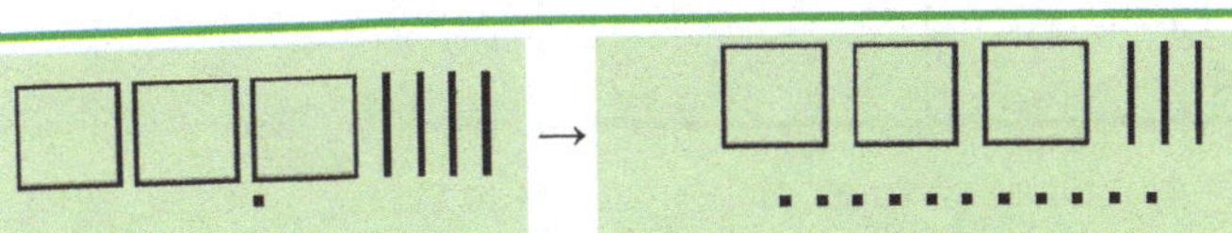

341 → 300 + 30 + 11

Take away 127. What is left? 214

b.
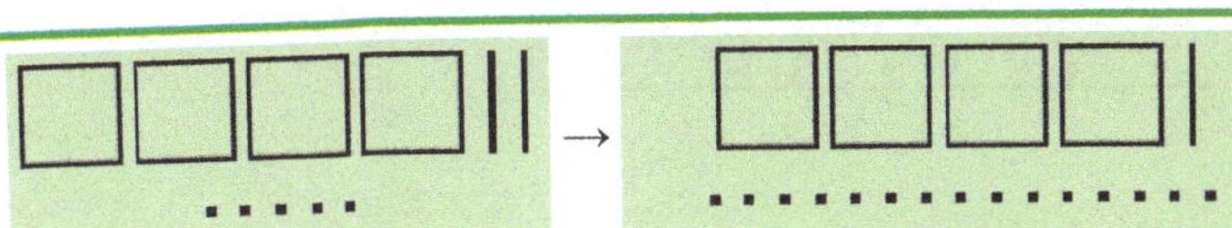

425 → 400 + 10 + 15

Take away 218. What is left? 207

c.
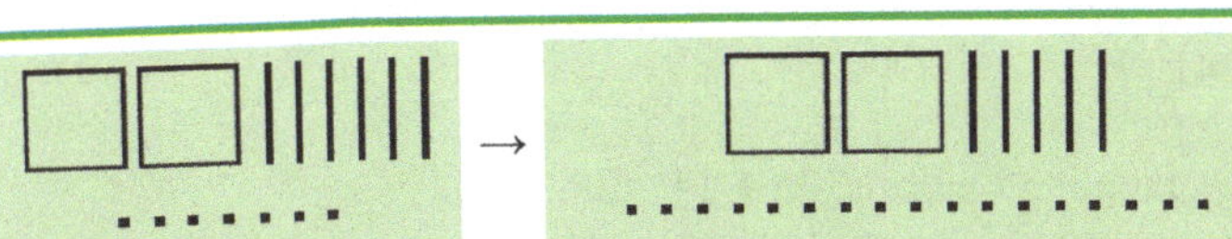

267 → 200 + 50 + 17

Take away 139. What is left? 128

2. a. 815 b. 467 c. 417

3. a. 618; 618 + 205 = 823 b. 216; 216 + 455 = 671
   c. 305; 305 + 115 = 420 d. 308; 308 + 208 = 516
   e. 106; 106 + 627 = 733 f. 378; 378 + 309 = 687

Regrouping One Ten as Ten Ones with 3-Digit Numbers, cont.

4.

a.	b.
120 + 120 = 240 121 + 119 = 240 122 + 118 = 240 123 + 117 = 240 124 + 116 = 240 125 + 115 = 240 126 + 114 = 240 127 + 113 = 240	1 + 119 = 120 2 + 118 = 120 3 + 117 = 120 4 + 116 = 120 5 + 115 = 120 6 + 114 = 120 7 + 113 = 120 8 + 112 = 120
Till you reach 240 + 0 = 240.	Till you reach 120 + 0 = 120.

5. a. There are 730 days in two years. 365 + 365 = 730
   b. They still have 104 km to travel. 173 – 69 = 104.
   c. He has 283 pages left to read. 581 – 235 = 346

Regrouping One Hundred as 10 Tens, p. 40

1. a. 160 b. 160
   c. 151 d. 80

2. a. 175
   b. 360
   c. 70
   d. 84
   e. 272

3.

It would be easy, except I cannot subtract _8_ tens from _4_ tens. So, I need to take one of the _9_ hundreds and break it down as tens. So, now I will have only _8_ hundreds but I will now get _14_ tens. Now I can subtract.

9 hundreds 4 tens 7 ones ⇨	8 hundreds	14 tens	7 ones
–	2 hundreds	8 tens	2 ones
	6 hundred	6 tens	5 ones

4. a. 184 b. 172

5. a. 780 b. 434 c. 92 d. 237
   e. 444 f. 552 g. 162 h. 372

6. a. The second book has 210 pages. 270 – 60 = 210
   b. Liz got 241 more points than Hannah. 433 – 192 = 241
   c. Hannah got 308 points. 215 + 93 = 308
   d. Micah caught 25 worms. 14 + 11 = 25. Together they caught 39 worms. 25 + 14 = 39

Puzzle corner:

4 4 5	6 8 4	9 8 1	9 6 0
– 1 5 3	– 3 5 8	– 2 5 5	– 4 5 5
2 9 2	3 2 6	7 2 6	5 0 5

Graphs and Problems, p. 44

1.

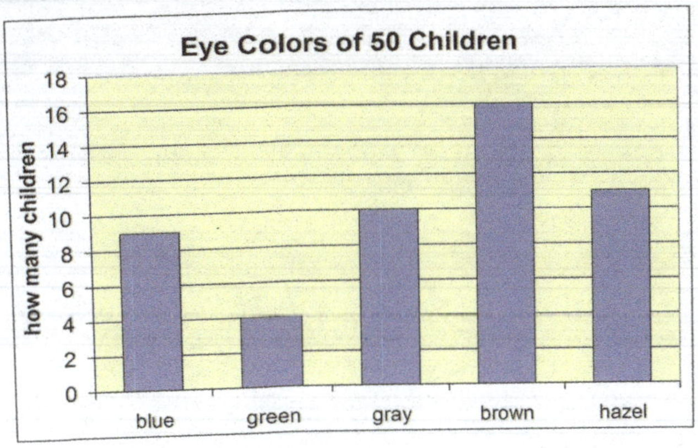

a. 27 children b. 12 children c. 41 children

2. a. No, together they have one less marble than Jim.
   b. You have read 58 pages. You have 37 pages left to read.

3. a. 90 birds b. 25 Australian animals
   c. 85 together d. 35 more e. 135 animals

4. a. 22 b. 61 c. 32

Euclid's Game, p. 46

1. No.
2. All the numbers that get marked are also even.
3. All the numbers that get marked will also be multiples of 5.
4. a. No. You cannot mark numbers that are greater than 90.
   b. Yes.
   c. No. You will only get to mark multiples of 10.
   d. Yes.

Review, p. 49

1. a. 692 b. 417 c. 718 d. 764

2. The three bicycles cost a total of $462. Add: 154 + 154 + 154 = 462.

3. a. 120, 820 b. 180, 290 c. 740, 550

4. It is 380 feet all of the way around. Add 120 + 70 + 120 + 70 = 380.

5. a. 34, 88 b. 15, 63 c. 35, 84 d. 723, 882 e. 165, 556 f. 304, 550

6. a. 8, 48 b. 8, 88 c. 5, 4 d. 9, 5 e. 16, 13 f. 43, 29

7. a. $\begin{array}{r} 2\,5\,4 \\ +\,4\,7\,7 \\ \hline 7\,3\,1 \end{array}$ b. $\begin{array}{r} 5\,8\,9 \\ +\,3\,2\,5 \\ \hline 9\,1\,4 \end{array}$ c. $\begin{array}{r} 2\,0\,6 \\ +\,6\,8\,6 \\ \hline 8\,9\,2 \end{array}$ d. $\begin{array}{r} 6\,8\,1 \\ +\,2\,1\,9 \\ \hline 9\,0\,0 \end{array}$

8. a. 39 people. 52 – 13 = 39
   b. 39 toys. 23 + 16 = 39
   c. 17 toys. 33 – 16 = 17
   d. 700 points. 465 + 145 + 90 = 700
   e. 40 jumping jacks. 26 + 14 = 40.

Review, cont.

9. a.

CHILD	POINTS
Charlie	15
Bill	24
Amy	28
Cindy	21
Sarah	19

b. 4 points c. 6 points

Puzzle corner:

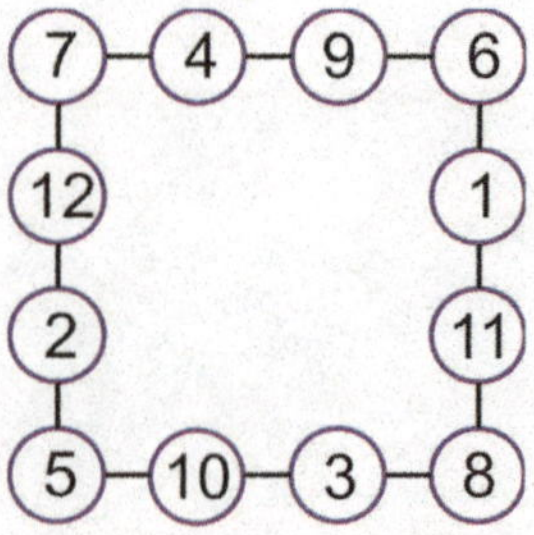

More from math MAMMOTH

Math Mammoth has a variety of resources to fit your needs. All are available as economical downloads, and most also as printed copies.

- **Math Mammoth Light Blue Series**
  A complete curriculum for grades 1-8. Each grade level includes two student worktexts (A and B), which contain all the instruction and exercises all in the same book, answer keys, tests, cumulative reviews, and a worksheet maker. International (all metric), Canadian, and South African versions are also available.
  https://www.MathMammoth.com/complete-curriculum
  https://www.MathMammoth.com/international/international
  https://www.MathMammoth.com/canada/
  https://www.MathMammoth.com/south_africa/

- **Math Mammoth Skills Review Workbooks**
  These workbooks are intended to be used alongside the Light Blue series full curriculum, and they provide additional review to the topics studied in the main curriculum, in a spiral manner.
  https://www.MathMammoth.com/skills_review_workbooks/

- **Math Mammoth Blue Series**
  Blue Series books are topical worktexts for grades 1-8, containing both instruction and exercises. They cover all elementary math topics from 1st through 8th grade and some for 8th grade. These books are not tied to grade levels, and are thus great for filling in gaps.
  https://www.MathMammoth.com/blue-series

- **Make It Real Learning**
  These activity workbooks concentrate on answering the question, "Where is math used in real life?" The series includes various workbooks for grades 3-12.
  https://www.MathMammoth.com/worksheets/mirl/

- **Review Workbooks**
  Workbooks for grades 1-8 that provide a comprehensive review of one grade level of math —for example, for review during school break or summer vacation.
  https://www.MathMammoth.com/review_workbooks/

Free gift!

- Receive over 350 free sample pages and worksheets from my books, plus other freebies:
  https://www.MathMammoth.com/free/

Lastly...

- Inspire4 is an inspirational website for the whole family I've been privileged to help with:
  https://www.inspire4.com

Made in the USA
Monee, IL
02 November 2026

67380100R00044